The Drone next door
A brief journey behind the eyes of the new-age H.a.w.k.

TENNYSON SAMUEL JOHN

Copyright © 2017 Tennyson Samuel John

All rights reserved.

ISBN: 9781521515648

DEDICATION

I would like to dedicate this book to my parents and express my sincere gratitude for their never-ending love, continual motivation and strong support extended at all times.

CONTENTS

	Abstract	i
1	Causal effect	8
2	Core science	12
3	Image Processing	46
4	Navigation Systems	52
5	System Validation	64
6	The near-future	87
7	Bibliography	91

ABSTRACT

In recent years, unmanned autonomous vehicles have been used in diverse applications because of their multifaceted capabilities. In most cases, the navigation systems for these vehicles are dependent on Global Positioning System (GPS) technology. Many applications of interest, however, entail operations in environments in which GPS is intermittent or completely denied. These applications include operations in complex urban or indoor environments as well as missions in adversarial environments where GPS might be denied using jamming technology.

This book investigates the development of vision-aided navigation algorithms that utilize processed images from a monocular camera as an alternative to GPS. The vision-aided navigation approach explored in this thesis entails defining a set of inertial landmarks, the locations of which are known within the environment, and employing image processing algorithms to detect these landmarks in image frames collected from an onboard monocular camera. These vision-based landmark measurements effectively serve as surrogate GPS measurements that can be incorporated into a navigation filter. Several image processing algorithms were considered for landmark detection and this thesis focuses in particular on two approaches: the continuous adaptive mean shift (CAMSHIFT) algorithm and the adaptable compressive (ADCOM) tracking algorithm. These algorithms are discussed in detail and applied for the detection and tracking of landmarks in monocular camera images. Navigation filters are then designed that employ sensor fusion of accelerometer and rate gyro data from an inertial measurement unit (IMU) with vision-based measurements of the centroids of one or more landmarks in the scene. These filters are tested in simulated navigation scenarios subject to varying levels of sensor and measurement noise and varying number of landmarks. Finally, conclusions and recommendations are provided regarding the implementation of this vision-aided navigation approach for autonomous vehicle navigation systems.

1 CAUSAL EFFECT

The void and the cure – A need to be addressed

The Global Positioning System (GPS) is used in multifarious systems as a position estimation sensor for vehicle navigation. Most inertial navigation systems (INS) are dependent on GPS to correct for drift errors that accumulate when processing inertial measurement unit (IMU) data (i.e., accelerometer and rate gyro data). For many scenarios of interest, however, such as UAV missions in urban environments, indoor operations, or space exploration missions, GPS availability may be intermittent, corrupted, or completely denied. As a result, there has been an increasing need for navigation solutions that do not depend on GPS.

The advent of computer vision and control theory in autonomous navigation applications, and the introduction of monocular camera systems on unmanned vehicles for the same purpose, has received growing interest as an alternative sensor to GPS systems. Therefore, vision-aided navigation systems represent a potentially important enabling technology for autonomous vehicle development. The diverse variety of available image processing algorithms serves as a primary reason for developing, implementing and testing two completely different tracking algorithms for advanced navigation applications with unmanned systems.

As Blake Anderson writes in his blog, "7-ways you could use a drone in construction projects", the vision systems in various forms have steered the increasing usage of drones in the ever growing construction and agricultural industries. Whether it's planning or inspection or analyzing or even improving safety, vision systems with high fidelity are in demand.

In the beginning – a blast from the past

Unmanned Aerial Vehicles (UAVs) constitute a research field that has been extensively explored in the last decade [1]. Earlier studies on autonomous vehicles have focused on modeling and identification, simulation, sensor integration, control design and fault analysis [2–5]. The effective use of vision odometry from high-end cameras interfaced with the onboard sensors as a suitable alternative method for autonomous vehicle navigation has been discussed frequently in recent years. These efforts were undertaken since the classical use of GPS to correct for drift in INS systems cannot sustain autonomous flight in GPS-denied environments [6, 7].

Early researchers in this field have overcome the problem of navigation and control in GPS-denied environments efficiently using generic techniques. Researchers have investigated laser devices such as scanning range finders and Light Detection and Ranging devices (LIDAR) for mapping the surrounding environment and use this information for navigation. However, these range sensors typically rely on the properties of the signal, but these estimates become too noisy and the accuracy is insufficient for estimating velocity for feedback control [8].

Vision-based tracking algorithms such as field estimation [9] and feature point tracking [10-11] and related research on computer vision techniques in UAVs have been applied to several applications. Camera sensors have a tremendous potential for localization, target identification and surface mapping applications since they provide data about features such as landmarks, corners and edges, and patterns in the environment, which can be used to infer information about vehicle motion and position [12]. One reference paper [13] proposed a visual odometry algorithm based on geometric homography. In this approach, vision data was used to compute a frame by frame odometry for a Simultaneous Localization and Mapping (SLAM) algorithm.

In earlier times, when the problem of instability arose due to camera mounting, in most cases contributing to addition of noise, poor navigation solutions were deemed responsible for causing a potentially destabilizing coupling between the navigation and control systems of the vehicle. The best solutions available were based on the use of camera based monocular SLAM (Simultaneous Localization and Mapping) techniques, which later were replaced by the incorporation of the inertial measurements from the IMU integrated with image processing data for navigation purposes. SLAM is considered to be a hybrid of two well-known problems: tracking and navigation through localization [13]. It is a mapping technique used to develop a map of an unknown environment or update a previous map thereby aiding navigation in the case of GPS-denied environments. The earliest use of this strategy was in the 1980's [14].

Vision-based methods have been proposed even in the context of autonomous landing at a landmark, as seen in [15]. In this work, inertial sensors are combined with a single camera and a specially designed landing pad as a landmark in order to be GPS independent. The problem of autonomously landing a UAV or helicopter in an unknown environment is discussed in [16]. In general, the degree of autonomy of an unmanned vehicle during navigation depends on factors such as the ability to cope with the unexpected loss of GPS signal and the ability to navigate using natural landmarks.

Several generic solutions for vision-based autonomous waypoint navigation and safe landing on unknown or known landmarks have been implemented using classical image processing techniques [17–19], and vision-based solutions have also been developed for the problem of collision awareness and avoidance [20-23].

The algorithms used earlier for extracting information from the camera were often computationally intensive and difficult to process taking into account lesser resources for data processing after extraction. To compensate for the loss of resolution and the inaccuracy of the data processing unit, a Kalman filter based navigation system integrated with an IMU and fused with a vision system were introduced to alleviate the computational burden by allowing frame to frame prediction of camera motion and object position in image frame, thereby helping to resolve scale ambiguity and improving overall accuracy.

The highlight reel – Scope of work

The two algorithms under the microscope or rather under the spotlight for a drone's object tracking capabilities for vehicle navigation in GPS-denied environments belong to two extremes of the image processing spectrum. The continuous adaptive mean shift (CAMshift) algorithm is highly data-dependent, thereby learning and training itself based on the changes in the environment and taking into consideration the factors affecting the change. The advanced compressive (ADCOM) algorithm is data independent in the sense that it makes use of the features extracted and classified based on positive and negative samples and continuously tracks these compressed features robustly. In general, these algorithms, which are developed towards the goal of landmark tracking for navigational purposes, employ visual cues based on shape, edges, color, intensity, corners, patterns, and center of mass of densest region for landmark identification and for estimating their position, thereby providing assistance for real-time tracking in environments where GPS data is inaccessible. In most cases, factors such as variation in lighting, shadows, occlusions, dynamic objects in the scene, and motion and vibration effects introduce an element of uncertainty, which may affect the robustness of an algorithm in real-time scenarios. The fidelity of the two algorithms and their tracking capabilities in severe environments pertaining to high risk applications is the core focus of this book which would pave way to a 360 degree view of what happens behind the screens when a drone takes off on a mission.

The primary reason for choosing these two algorithms is that from previous work it was understood that algorithms having the same fundamental base or mode of operation have been tested for object detection. This work gives an option for the user to choose between two algorithms belonging to two highly different operational modes for suitable applications in object tracking. Also, the data obtained from the post-processing phase can be used for navigation of unmanned vehicles in GPS-denied environments.

The study of the efficiency and versatility of the two algorithms is validated by experiments conducted in various conditions for navigation purposes. The developed navigation filter would provide a solution for vehicle navigation in remote areas. This would eventually aid the domain of vision-aided navigation using landmark feature tracking in GPS-denied environments, thereby expanding the scope to newer, flexible methods for commercial and military applications.

An intelligible roadmap - The scientific objectives

This book emphasizes some unique image processing methods, which are adopted for navigation in GPS-denied environments by making use of processed data from high-end cameras as a replacement for GPS navigation measurements. A key objective of the research is to provide an in-depth analysis of the emerging tracking algorithms, which are put to test in specific applications such as UAV vision-based object tracking under various conditions. An extended Kalman filter is developed to process the real-time data provided by the image processing algorithms. The primary technical objectives are:

1. Generate a detailed vision data set using an Unmanned Ground Vehicle (UGV) and an Unmanned Aerial Vehicle (UAV) equipped with video cameras for the purposes of 3D landmark recognition.

2. Derive and analyze robust image processing techniques for identifying known landmarks in the environment from the vision data using video from monocular cameras.

3. Based on the landmarks that are identified within the scene, derive navigation laws/filters to integrate the landmark measurements with IMU measurements to compute a GPS- denied navigation solution.

2   CORE SCIENCE

The concepts of object recognition, detection and tracking have been explored in the field of computer vision starting in the early 1980's. The earliest use of these techniques was tested in a project using an **IMSAI 8080** microcomputer, Altair computers and a Cromemco Cyclops camera. The software used was written in assembly language, with some basic running on under CPM (a control program for microcomputers), which paved way for the floppy disk to be shared by multiple computers. The camera was programed to recognize, detect and track a ball even though it functioned under a low **32 x 32 bit** resolution. The ball was made black for high contrast and fast positioning determination. The maze surface used was represented as a ball position, speed and direction vector map/matrix. This entire process used 3 computers: one for ball position and speed through the camera data, one for stepping motor control and one for the coordination of the ball position and speed in the vector field.

As the name suggests, object recognition involves the positive identification of an object based on one or more properties of the object under consideration such as specific features, edges, corners, and color intensity. Object detection is similar in principle to object recognition, where the target in a frame is identified and compared to either a specific shape or pattern or a template, and then labeled or segmented as observed. This chapter emphasizes two important detection and tracking algorithms that were developed in order to suit the core purpose of this research. These algorithms can be implemented as real-time tracking algorithms in the flight computer of a fixed wing UAV or a quadrotor for the purpose of vision-aided navigation. Also, some secondary algorithms, which could be used alongside the primary algorithms, are discussed. The two primary object detection and tracking algorithms considered in this book are the continuous-adaptive mean shift (CAMshift) algorithm and the adaptable compressive tracking algorithm (ADCOM).

2.1 CAMshift Algorithm

The Continuously Adaptive Mean Shift CAMshift algorithm is an advancement of the mean-shift tracking algorithm, which provides a robust algorithm for object detection and tracking. The tracking is predominantly dependent on properties such as color, edge, and texture. The CAMshift algorithm reduces the tracking error by continuously changing the search window based on target properties in cases where the scaling of the image is improper or the search window is either too small or too large. Basically, the algorithm adapts to changes in the color, scaling, orientation, and texture of the image during the target search and localization processes. A key enhancement of the CAMshift algorithm over the mean shift algorithm is that CAMshift is designed specifically for dynamically changing distributions while the mean shift algorithm is designed for static distributions.

The CAMshift algorithm is composed of the following basic steps:

- The target to be tracked is chosen at first either by providing its position in the first frame or manually selecting it.
- A histogram of the hue signal from the complete image is generated.
- The region of interest is increased and the zeroth and first moments of the image, which correspond to the area and mean of the probability density function, respectively, are calculated.
- The mean shift algorithm is iterated until convergence occurs.
- The centroid calculated used to center the search window for the next frame and the zeroth moment is used to recalculate the image size and window size.
- The mean shift algorithm is iterated continuously until all frames are computed (i.e., the mean location moves less than a preset threshold).

When tracking a colored object from a video sequence, CAMshift operates on a color probability distribution image derived from color histograms. CAMshift calculates the centroid of the 2D color probability distribution within its 2D window of calculation, re-centers the window, and then calculates the area for the next window size. Thus, it is not necessary to calculate the color probability distribution over the whole image, but instead the calculation of the distribution can be restricted to a smaller image region surrounding the current CAMshift window [25].

The essence of the CAMshift algorithm lies in the idea of region matching, a process to find correspondences between the image elements from two image frames with different viewpoints. The match criterion is the similarity based out of the color probability distribution, which is continuously changing based on the changes in the image elements in consecutive frames.

The mean shift algorithm determines the convergence from the initial search window location and scales the best match based on the color-histogram similarity. The location is updated by applying the mean shift algorithm across each pixel location and scaling is performed to find the densest region of similarity to the target's probability density function. This similarity refers to the Bhattacharyya distance d, represented by its coefficient r.

$$r[\tilde{p}(y),\tilde{q}] = \sum_{u=1}^{m} \sqrt{\tilde{p}_u(y)\tilde{q}_u}$$

$$d[\tilde{p}(y),\tilde{q}] = \sqrt{1 - r[\tilde{p}(y),\tilde{q}]}$$

(2.1)

The algorithm makes use of the mean shift theory in order to determine the probability density histogram over successively changing frames. In equation 2.1, 'p' and 'q' represent the pixel range in 2D for the target and candidate model. The Bhattacharyya distance show the similarity between the candidate and target histograms under consideration.

2.1.1 Mean Shift Theory

The mean shift algorithm is a robust, nonparametric technique that ascends the gradient of the probability distribution to find the mode or peak of this distribution [26]. The earliest usages of this algorithm were seen in mode seeking, particle filtering, and kernel-based object tracking applications [27]. In order to apply the algorithm, a region of interest (or window) is first selected, then the centroid of this ROI is obtained for every frame, and finally the centroids for all the frames are linked to provide object tracking.

To explain the algorithm more clearly, a collection of points or scattered pixels is considered in 2D and the densest region in the collection of points is found. An initial ROI is selected and its centroid is noted. The center of mass for the group of points is then calculated. This is similar to the centroid for the weight of all the points in the ROI, which may not be the same as the centroid of the ROI.

The mean shift vector is found as the shift or distance between the initial estimate (initial mean of the data points) and the new center of mass. The centroid is then shifted to the new center of mass. This procedure of finding the centroid and center of mass and the mean shift vector is repeated iteratively until the centroid and center of mass overlap, indicating the densest region. The mean shift vector M_h is represented as:

$$M_h(y_0) = \left[\frac{\sum_{i=1}^{n_x} W_i(y_0) x_i}{\sum_{i=1}^{n_x} W_i(y_0)} \right] - y_0 \qquad (2.2)$$

Where x_i are the data points as 2D vectors, W_i is the weight for each data point (dependent on the distance from the initial mean location to that particular data point), n_x is the total number of data points and y_0 is the initial estimate or mean location. The process is performed iteratively until the mean shift vector becomes zero, indicating that the densest region has been reached.

The mean shift vector [28] always points towards the direction of maximum increase in density. The denominator of the above equation represents the added weights for all data points, which is referred to as the "kernel".

Some noteworthy properties of the mean shift algorithm are:

- The mean shift vector has the direction of the gradient of the density estimate.
- The mean shift vector is computed iteratively to obtain the maximum density in the local neighborhood.
- The mean shift algorithm is an effective means of finding modes or peaks in a set of data points in a distribution manifesting an underlying probability density function (PDF).

It can be noted that, given a set of data, mean shift helps to analyze the distribution that is nonparametric. Parametric distributions, such as Gaussian or exponential distributions, can be represented in terms of analytical formulas. Nonparametric distributions describe distributions in which a Gaussian or a mixed-multiple Gaussian curve cannot be used to fit the data points. Computing the gradient of the nonparametric distribution using the mean shift is better than the nonparametric density estimation without the gradient. In this process, the mode is found as the peak in the distribution, and the mode corresponds to the gradient. The height of the distribution in a histogram is proportional to the number of data points.

A taller histogram suggests a denser region with more points. This can be also indicated using the Bhattacharyya coefficient. The kernel density estimation for the nonparametric data results in finding the probability density function. The weights are assigned based on the different types of kernels such as Gaussian, uniform and Epanechnikov. Each of these kernels has a specific profile associated with it. Equation (2.3) represents the probability density function arising from a radially symmetric kernel whose profile is given as 'k'. This equation leads to a relationship with the mean shift vector.

$$P(x) = \frac{c}{n} \sum_{i=1}^{n} k(\|x - x_i\|^2) \qquad (2.3)$$

The gradient of the probability function given above would lead to equation (2.4) which can be directly related to mean shift vector when compared to equation (2.3).

$$\nabla P(x) = \frac{c}{n} \sum_{i=1}^{n} g_i \left[\frac{\sum_{i=1}^{n} x_i g_i}{\sum_{i=1}^{n} g_i} - x \right] \quad ; \text{(here, } g_i = g(\|x - x_i\|^2) = \nabla k(\|x - x_i\|^2) = \nabla k_i) \quad (2.4)$$

It can be observed from equation (2.3) and (2.4) that the mean shift vector is one of the terms in the gradient of probability function. In other words, a relationship exists between the two factors which is given in the equation (2.5).

$$\text{Mean Shift Vector, } M_h(x) = \frac{\nabla P(x)}{\left[\frac{c}{n}\right] \sum_{i=1}^{n} g_i} \quad ; \text{(where } g_i \text{ represents the weights)} \quad (2.5)$$

The mean shift algorithm incorporates some crucial stages that are employed in the proper execution of the CAMshift algorithm. These stages are given as follows:

- Choose a search window and an initial location of the search window.
- Compute the mean location (centroid) in the search window by finding the zeroth moment and then the first moment for x and y.
- Center the search window at the computed mean location.
- The above steps are carried out until a convergence occurs when the mean reaches zero and the centroid converges with the center of mass (densest region).

Figure (2.1) depicts the mean shift process as the convergence occurs at the densest region.

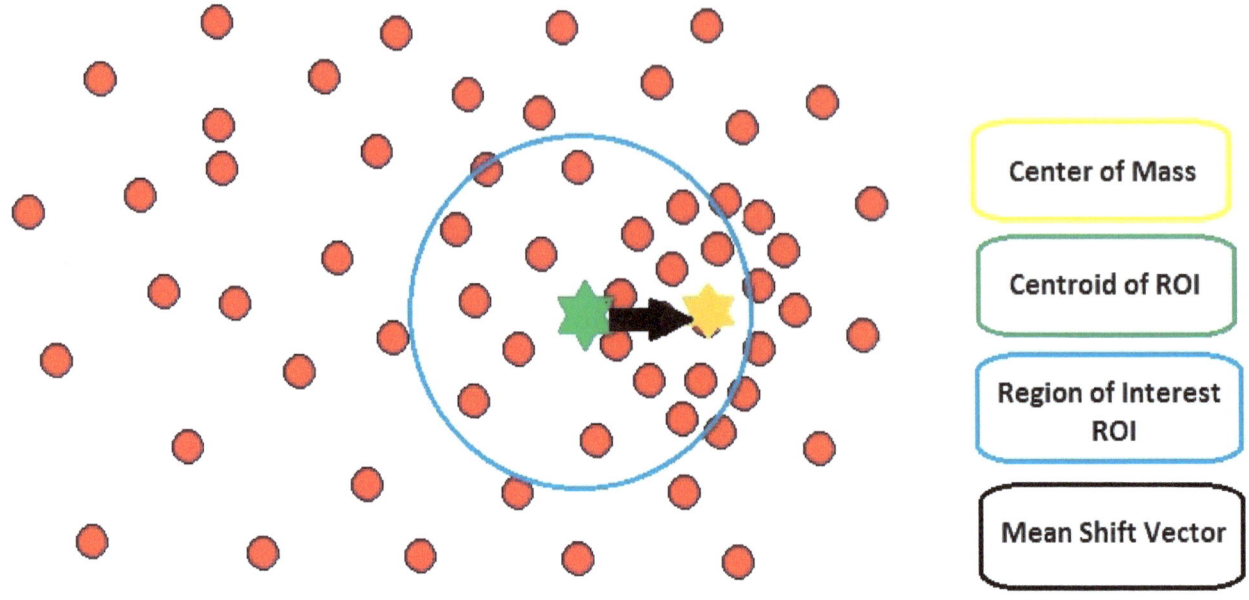

Figure 2.1 Mean Shift Convergence

2.1.2 Mass center calculation

The mean shift convergence criteria states that the mean shift component is implemented by continually computing new values of the mean or centroid location (x_c, y_c) for the window position computed in the previous frame until there is no significant shift in position. The maximum number of Mean Shift iterations is usually taken to be 10-20 iterations. Since sub-pixel accuracy cannot be visually observed, a minimum shift of one pixel in either of the x and y directions is selected as the convergence criteria. Furthermore, the algorithm must terminate in the case where M_{00} (the first moment) is zero, which corresponds to a window consisting entirely of zero intensity. For the CAMshift process the zeroth, and first moments are calculated initially as a part of mean shift approach in order to compute the centroid for the tracked object. They can be given as the following mass center calculations:

a) Compute the zeroth moment

$$M_{00} = \sum_x \sum_y I(x, y) \tag{2.6}$$

Where I(x, y) represents the intensity of discrete probability image at (x, y) within the search window.

b) Find the first moment for x and y

$$M_{10} = \sum_x \sum_y x I(x, y)$$
$$M_{01} = \sum_x \sum_y y I(x, y) \tag{2.7}$$

c) Compute the mean search window location

$$x_c = \frac{M_{10}}{M_{00}}; \quad y_c = \frac{M_{01}}{M_{00}} \tag{2.8}$$

CAMshift is designed for dynamically changing probability distributions. These occur when a tracked object moves so that the size and location of the probability distribution changes in time at a significant rate. The CAMshift algorithm adjusts the search window size in the course of its operation. The initial window size can be set at any reasonable value. CAMshift uses the zeroth and first moment to continuously adapt its window size within or over each video frame. The window radius, or height and width, is set to a function of the zeroth moment found during the continuous search. The algorithm is then performed using any initial nonzero window size.

The motivation for the CAMshift algorithm, which is a modification of the mean shift theory, is that the mean shift algorithm fails as a tracker when subjected to adverse real-time conditions. A window size that works at one distribution scale is not suitable for another scale as the color object moves towards or away from the camera.

The mean shift algorithm makes use of a single probability density function for the entire tracking process. Therefore, the algorithm may fail when subjected to camera motion, occlusions or noise in the form of distractors, inclusive of a large sized or a small sized search window.

The flowchart in Figure (2.2) describes the CAMshift algorithm with the implementation of the mean shift theory. The algorithm continuously adapts the search window based on the scaling and orientation of the tracked object.

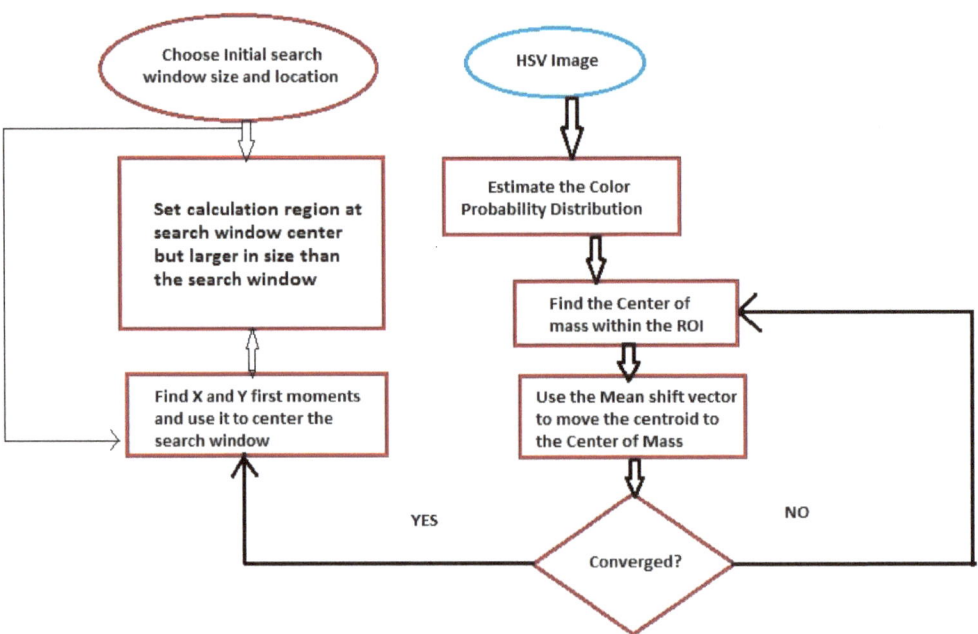

Figure .2.2 CAMshift Implementation

2.1.3 Histogram Back-projection

The probability distribution function (PDF) may be determined using any method that matches a particular pixel value to a probability distribution of a landmark or the surrounding [29]. A common method used for this calculation is the histogram back-projection.

First, the PDF is generated by obtaining an initial histogram using the first step of the CAMshift algorithm process from the initial search window (ROI) of the filtered image.

The histogram generated makes use of the hue channel in HSV color space. As discussed by Bradski, the Hue Saturation Value (HSV) color system corresponds to projecting standard Red, Green, Blue (RGB) color space along its principle diagonal from white to black (shown in Figure (2.3a)), which results in the hex cone shown in Figure (2.3b). Descending the V axis in Figure (2.3b) gives smaller hex cones corresponding to smaller (darker) RGB subcubes in Figure (2.3a). The unique quality of HSV space separates out hue (color) from saturation (color concentration) and brightness. The color models are then created by computing 1D histograms from the H (hue) channel in HSV space.

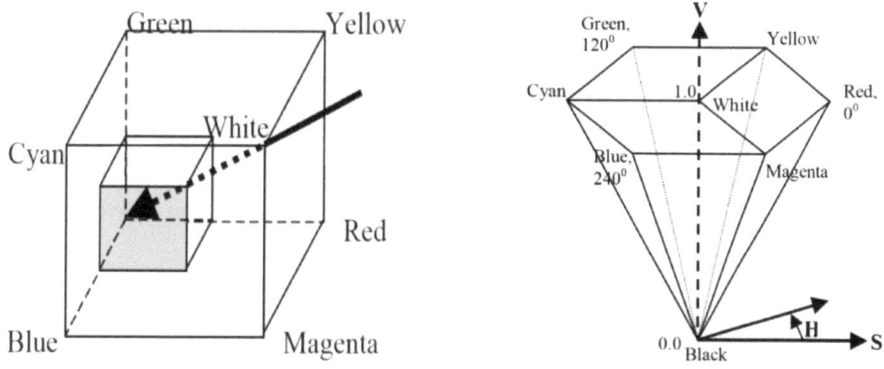

Figure 2.3(a) RGB color cube 2.3(b) HSV color system

A noteworthy consideration when using real cameras with discrete pixel values is that a problem can occur when using HSV space as can be seen in Figure 2.3(b). When brightness is low (V near 0), saturation is also low (S near 0). Hue then becomes quite noisy, since in such a small hex cone, any small changes in RGB values would not be adequately represented by discrete hue pixels. This then leads to large deviation in the hue values, causing an erroneous and noisy histogram.

To overcome this problem, hue pixels that have very low corresponding brightness values are neglected. This means that, for very dim scenes, the camera must auto-adjust or be adjusted for more brightness in order to track effectively. With sunlight and with bright white colors, an upper threshold can be applied to ignore hue pixels with corresponding high brightness. At very low saturation, hue is not defined, so hue pixels that have very low corresponding saturation are also ignored.

The 1D histogram obtained initially is quantized into bins, which lessens the computational and spatial complexity and allows similar color values to be clustered together [30]. The histogram bins are then scaled between the minimum and maximum probability image intensities using Equation (2.9):

$$\left\{ P_u = \min\left(\frac{255}{\max(q_u)} q_u, 255 \right) \right\}_{u-1....m} \tag{2.9}$$

Here 'm' denotes the number of bins, P_u and q_u represent the discrete pixel range of 2D probability and histogram respectively. Histogram back-projection is a primitive operation that associates the pixel values in the image with the value of the corresponding histogram bin by allowing the re-application of the histogram values generated already on the current frame. The back-projection of the target histogram with any consecutive frame generates a probability image where the value of each pixel characterizes the probability that the input pixel belongs to the histogram that was used. This is accommodated by rescaling the probability (histogram values) to a new range, where pixels with highest probability of being in the sample histogram map as visible intensities in the 2D histogram back-projection image. The CAMshift algorithm deals with image problems that frequently occur during colored object recognition and tracking such as irregular object motion due to perspective, image noise, distractors, and occlusions.

The algorithm constantly re-scales itself in a way that naturally fits the structure of the image data. Based on the potential velocity and acceleration of the object and the scaling based on its distance to the camera, the size of its color probability distribution in the HSV plane is scaled accordingly. In this manner, when objects closer to the camera move rapidly in the image plane, their color distribution occupies a larger area.

In this situation, the search window size gets scaled larger, hence capturing larger movements. When objects are distant, the color distribution is small so the search window size is also small, but distant objects are slower to traverse the video scene. This natural adaptation to distribution scale and translation without predictive filters or a parametric distribution helps further the computational savings and serves as a built-in remedy to the problem of erratic object motion and sudden changes in the camera motion. This process ignores the distribution outliers and substantially reduces occlusions and noise; thus tracking parameters need not be smoothed, filtered, or sampled resulting in robust noise tolerance.

The robustness of CAMshift with respect to noise, transient occlusions, and distractors mostly depends on the search window matching the size of the object being tracked. It has been shown that it is better to err on the side of the search window being too small than too large. The search window size depends on the function of the zeroth moment M^{00}. To indirectly control the search window size, we adjust the color histogram up or down by a constant, truncating at zero or saturating at the maximum pixel value. The adjustment affects the pixel values in the color probability distribution image which affects M^{00} and hence the window size. For an 8-bit hue signal, the histogram is adjusted by 20 to 80 (maximum being 255), which tends to reduce the search window size to just within the size of the tracked object, hence eliminating noise. HSV brightness and saturation thresholds are employed in cases where hue is not well defined for very low or high brightness or low saturation. Low and high thresholds are set from 10 – 20% of the maximum pixel value.

It can be noted that the Kalman filtering process, or any other smoothening process typically used in object tracking, is an optimal estimation method with the criterion of minimal error covariance. It has the advantages of low calculation scale and real-time performance. However, the standard Kalman filter employs a linear Gaussian state space model, which may be not consistent with the motion of the tracked object in the real world. In comparison, an improved object tracking algorithm such as the CAMshift algorithm has discrete advantages in that it does not make assumptions regarding the motion of the camera and can provide improved tracking.

2.1.4 Effect of Scaling and orientation features in CAMshift algorithm

The use of moments to determine the scale and orientation of a probability distribution of a tracked object was first described in 1986 [31] [37] and later in the late 1990's by Freeman and Bradski. Here, the scale and orientation of the target are estimated using the moment features of the weight image. Therefore, once the weight image is properly calculated, it can lead to accurate moment features and consequently good estimates of changes in the target.

The scale and orientation can be well estimated using the probability density function together with the moment features of the weight image [32]. Firstly, in order to estimate the weighted area of the target in the target candidate region, the zeroth moment (M_{00}) is taken into account. The Bhattacharya's coefficient is an indicator showing the similarity between the target model and the target candidate model. Higher this coefficient means more features from the target are present than the background [33] [34] [35]. Also, it also reveals a lower estimation error considering the zeroth moment. Therefore Bhattacharya coefficient is a reliable indicator considering its effect on the zeroth moment and thereby the target area. This can be given by the following equation:

$$A = c(r)M_{00}, \text{Where } A \text{ is the Area and } M_{00} \text{ is the Zeroth Moment}$$
$$\text{Where } c(r) = r = 1 \text{ in case of a linear function (or)}$$
$$c(r) = \exp\left(\frac{r-1}{sig}\right), 0 \le r \le 1 \text{ in other cases} \quad (2.10)$$

Based on the experimental results for this thesis, the linear function is put to use and tested successfully. Also, in other experiments conducted previously where the exponential function was used the value of "**sig**" ranges between 1 and 2 for robust tracking results adaptive to the video content.

In order to determine the orientation of the major axis and the scale of the distribution, an equivalent rectangle is found that has the same moments as those measured from the 2D probability distribution image. Two methods are suggested and compared for calculating the orientation and scale of the object.

The first method is the conventional type, which uses the following first and second moment expressions to compute the scaling and orientation estimates:

$$M_{11} = \sum_x \sum_y xy I(x, y)$$
$$M_{20} = \sum_x \sum_y x^2 I(x, y) \quad\quad (2.11)$$
$$M_{02} = \sum_x \sum_y y^2 I(x, y)$$

The first two eigenvalues, the major length and width of the probability distribution, are calculated in closed form as follows with the help of the first and second order central moments given below.

$$m_{11} = 2\left(\frac{M_{11}}{M_{00}} - (x_c y_c)\right),$$
$$m_{20} = \left(\frac{M_{20}}{M_{00}} - x_c^2\right); m_{02} = \left(\frac{M_{02}}{M_{00}} - y_c^2\right) \quad\quad (2.12)$$

From the above equations the minor and major semi-axes w and l of the probability distribution can be determined as follows.

$$Major\ semi-axes, l = \sqrt{\frac{(m_{20}+m_{02})+\sqrt{m_{11}^2+(m_{20}-m_{02})^2}}{2}}$$
$$Minor\ semi-axes, w = \sqrt{\frac{(m_{20}+m_{02})-\sqrt{m_{11}^2+(m_{20}-m_{02})^2}}{2}} \qquad (2.13)$$

The object orientation or the head roll, in case of face tracking, or major axis inclination can be found making use of the central moments as well. In most cases the CAMshift helps in tracking four degrees of freedom. The orientation is given by:

$$Theta = \frac{1}{2}\arctan\left(\frac{m_{11}}{m_{20}-m_{02}}\right) \qquad (2.14)$$

By using this method, it can be seen that the CAMshift algorithm can be used for general-purpose object tracking using a background-weighted histogram and arbitrary quantized color features of the target, but the scaling with orientation values would not be as accurate as the second method using covariance matrices.

The second method entails the formation of a covariance matrix using the second order central moments. The covariance matrix given below is estimated in order to obtain the width, height and orientation of the target.

$$Cov = \begin{pmatrix} m_{20} & m_{11} \\ m_{11} & m_{02} \end{pmatrix} \qquad (2.15)$$

Therefore, by making use of the estimated area A from equation (2.5) and the central moment features, the width, height and orientation are computed. This is specifically done using the singular value decomposition (SVD) process [38]. The SVD method implies that

$$Cov = V \times S \times V^T = \begin{pmatrix} v_{11} & v_{12} \\ v_{21} & v_{22} \end{pmatrix} \times \begin{pmatrix} l_1^2 & 0 \\ 0 & l_2^2 \end{pmatrix} \times \begin{pmatrix} v_{11} & v_{12} \\ v_{21} & v_{22} \end{pmatrix}^T \qquad (2.16)$$

where l_1^2 and l_2^2 are eigenvalues of the matrix Cov

Also, the orientation of the two main axes, pitch and roll estimation, is given by two vectors of matrix V given as (v_{11}, v_{21}) and (v_{12}, v_{22}) respectively. A noteworthy comparison between the two methods is that the values of l and w correspond to l_1 and l_2 respectively. But the values calculated in the second method are more accurate.

An advancement to the above method has been developed to compute a more accurate width and height estimation. In some cases, the target is represented in the form of an ellipse for which the lengths of the semi-major axis and semi-minor axis are denoted by a and b, respectively.

Theoretically it is proven that 'a' and 'b' can be related to l_1 and l_2 as follows:

$$\frac{l_1}{l_2} \approx \frac{a}{b} \Rightarrow a = kl_1 \text{ and } b = kl_2$$

where 'k' is the scale factor

By equating the known area of the target representation, either a rectangle or an ellipse, with the area A estimated from Equation (2.10), the following expressions are obtained:

$$k = \begin{cases} \sqrt{A/(l_1 l_2)} & \text{for RECTANGULAR TARGET} \\ \sqrt{A/(3.14 \times l_1 l_2)} & \text{for ELLIPTICAL TARGET} \end{cases}$$

$$a = \begin{cases} \sqrt{l_1 A/(l_2)} & \text{for RECTANGULAR TARGET} \\ \sqrt{l_1 A/(3.14 \times l_2)} & \text{for ELLIPTICAL TARGET} \end{cases} \quad (2.17)$$

$$b = \begin{cases} \sqrt{l_2 A/(l_1)} & \text{for RECTANGULAR TARGET} \\ \sqrt{l_1 A/(l_2)} & \text{for ELLIPTICAL TARGET} \end{cases}$$

The adjusted covariance matrix with the above implementation in place is given as below:

$$Cov = V \times S_1 \times V^T = \begin{pmatrix} v_{11} & v_{12} \\ v_{21} & v_{22} \end{pmatrix} \times \begin{pmatrix} a^2 & 0 \\ 0 & b^2 \end{pmatrix} \times \begin{pmatrix} v_{11} & v_{12} \\ v_{21} & v_{22} \end{pmatrix}^T$$

A fundamental but essential distinguishing factor between algorithms that estimate iteratively the covariance matrix for consecutive frames based on mean shift tracking and the above method is that the latter algorithm efficiently makes use of the area of the target representation with the covariance matrix to estimate the width, height and orientation of the object. This algorithm can robustly estimate the scale and orientation changes of the target under the CAMshift framework. The CAMshift algorithm with augmented scaling and orientation estimation techniques provides a simple and effective method to estimate the attitude and scaling features of the target as well as successfully tracking the object using the adaptive feature incorporated in the algorithm.

2.2 Adaptable (Advanced) Compressive (ADCOM) Tracking Algorithm

Compressive sensing theory uses a unique set of discrete audio or video signals, which are adequately, sensed using far fewer measurements than the dimension of the ambient space in which they appear. This technique constantly maintains the appearance of the model in the case of an image or maintains the quality of the audio signal without compromising the structure of the signal. The signal under consideration is accurately obtained from the data collected during the sensing process. There is no loss of data or addition of noise when this technique is implemented.

2.2.1 Background Subtraction Based Tracker

From previously used methods, it can be seen that large amounts of data can deteriorate the tracking functionality. One of the most intuitive applications of compressive sensing in visual tracking is the modification of background subtraction [39] such that it is able to operate on compressive measurements. Background subtraction aims to differentiate the object-containing foreground from the background.

This process not only helps to localize objects, but also reduces the amount of data that must be processed at later stages of tracking [40] [41].

Usually, the foreground information occupies a sparse spatial support compared to the background and may be caused by the motion and the appearance change of objects within the scene. By obtaining the object silhouettes on a single image plane or multiple image planes, a background subtraction algorithm is conventionally performed. However, traditional background subtraction techniques require that the full image be available before the process can begin. Background subtraction techniques may require complicated density estimates for each pixel, which become burdensome in the presence of high-resolution imagery.

2.2.2 Particle Filter Based L1 Tracker

The concept of sparse representation has been recently used in the L1 tracker where an object is modeled by a sparse coding process with a given dictionary of target and trivial templates. The algorithm seeks to find the best candidate that negates the reconstruction error using only this sparse linear combination of target and trivial templates. However, the computational complexity of this tracker is rather high, demanding high processing speeds, thereby limiting its applications in real-time experiments.

The L1 tracker and its extensions have been developed in the particle filter (PF) framework. A common problem arising in vision tracking is to estimate the posterior probability distribution. By definition, the posterior probability is the probability distribution of an unknown quantity, treated as a random variable, conditional on the relevant evidence or ground truth taken into account from a practical real-time experiment. In Bayesian statistics, the particle filter framework is a sequential Monte Carlo method, using samples of the conditional distribution in order to approximate it and thus the desired estimates. The L1 tracker makes use of a technique known as sequential importance sampling in the form of a bootstrap filter.

The sampling technique involves three stages. First, the samples are drawn from the known prior distribution. Second, a prediction step involving generation of candidate samples is followed by calculating importance weights based on the observation, which are later normalized. For each particle, its representation is computed independently by solving a constrained L1 minimization problem with non-negativity constraints.

Finally, the filter enters the selection step where samples are generated from a discrete distribution over the candidate particles. In order to adapt to the appearance and structural changes of the tracked object, the template is updated depending on both the weights assigned to each template and the similarity between templates and current estimation of target candidate.

2.2.3 Fundamental Types of Online Tracking Methods

The ADCOM algorithm used in this thesis is derived from the three types of online tracking methods discussed below. This discussion highlights the vast difference, flexibility and versatility of the ADCOM algorithm when compared to the other available visual tracking methods [42].

Generative Method

The generative method involves learning a model to represent the object and then using the model to search for the image with minimal reconstruction error. The tracking problem is formulated as searching for the regions that are most similar to the tracked targets. These are based on either template matching or subspace models. The continuously adaptive mean shift algorithm (CAMshift) is an example of this method. These methods consider the color probability distribution or the color histogram in their tracking approach.

Discriminative Method

The discriminative method poses the tracking problem as a binary classification task in order to find the decision boundary for separating the object from the background; hence it subtracts the background information or totally ignores the background data, thereby drawing an imaginary boundary between the features that best discriminate between object and background data. Online updating of the learned target is a common feature of this approach. Mostly, the negative samples are eliminated and only the positive features are learned and tracked. Drift is a major issue to be addressed when implementing these methods. The tracking, learning and detecting (TLD) algorithm is an apt example of the discriminative method.

Collaborative Method

Collaborative methods involve a semi-supervised learning approach in which positive and negative samples are selected via an online classifier with structural constraints. Both foreground and background information are used in the tracking process. While most tracking algorithms work on the premise that the object appearance or environmental lighting conditions or intensity do not significantly change as consecutive frames are processed, the ADCOM algorithm belongs to a category of online tracking algorithms that account for the appearance variation of the target and background, thereby facilitating the tracking task in various circumstances. The compressed samples of foreground targets and the background are obtained using the same sparse measurement matrix. The tracking task is formulated by classifying the positive and negative samples via a naive Bayes classifier under the Bayesian framework.

2.2.4 Need for a Stable, Robust and Efficient Algorithm

The most challenging task is to develop effective appearance models for robust and optimal object tracking solutions due to weakening factors like pose variation, illumination fluctuation, occlusion and motion blur [43] [44]. A current feature of online tracking algorithms often updates the appearance model with samples from observations in previous or recent frames. There are two main deviations to this feature. First, while these adaptive appearance models are data-dependent, there are very few data that can be properly utilized by these algorithms at the outset. Second, online tracking algorithms often encounter the problem of drift. This means that algorithms that adapt to their surrounding conditions work with misaligned samples that overlap each other and accumulate causing the degradation of the appearance model.

The solution proposed is in the form of an algorithm with an appearance model based on features extracted from multi-scale image feature space with data-independent features. Compressive tracking [45] is introduced to help alleviate some of the challenges associated with performing classical tracking in the presence of large amounts of data. The amount of data that the system must handle can be drastically reduced using this technique. The algorithm employs non-adaptive random projections that preserve the structure of the image feature space of objects. A very sparse measurement matrix is adopted to efficiently extract the features for the appearance model. In this algorithm, two important factors to be considered are the non-adaptive advanced random projections and the sparse measurement matrix. These two features are substantial to compressive sensing behavior and hence employed in the object tracking domain. This algorithm has been previously tested in environments where a static background is processed using data-independent features.

2.2.5 Random Projection

The concept of using random projection for compressive tracking revolves around a core foundation built on the Johnson-Lindenstrauss Lemma principle [46], which needs to be satisfied in order to apply random projection to compress a matrix. It states that if a set of points in a high-dimensional space is given, then these points can be projected into a much lower-dimensional random subspace that is independent of the original dimension, and with high probability preserve much of its structure in terms of its inter-point distances (Euclidean distance) and angles.

Let $A \in R^{n \times D}$ be our n points in D dimensions [47]. The method multiplies A, which corresponds to the image space, by a random matrix $C \in R^{D \times k}$, reducing the D dimensions down to just 'k' for speeding up the computation. C matrix is the random measurement matrix which typically consists of entries of standard normal N (0, 1). The projected data matrix which is of a lower dimensional image space is given by,

$$B = A * C \in R^{n \times k} \qquad (2.18)$$

If the random measurement matrix C in (2.18) satisfies the Johnson-Lindenstrauss lemma, the A matrix can be reconstructed with minimum error from B with high probability and robustness if and only if the A matrix is compressive in nature, such as in an audio or image vector space. In this way, all the information in B is preserved by the matrix A in a higher dimension. This theoretical foundation enables the linear analysis of the high-dimensional signals via low-dimensional random projections. For an efficient projection, a very sparse measurement matrix needs to satisfy the restricted isometry property (RIP).

According to the restricted isometry property [48], consider the random matrix C which is of size $D \times k$ is said to satisfy the RIP with the restricted isometry constant R (m, D, k; C) if, for every $A \in X^k(m) := \{A \in R^k : \|A\|_0 \leq m\}$,

$$R(m, D, k; C) := \min_{y \geq 0} y \text{ subject to } (1-y)\|A\|_2^2 \leq \|AC\|_2^2 \leq (1+y)\|A\|_2^2 \quad (2.19)$$

The RIP constant R(m,D,k;C) is the maximum distance from 1 of all the eigenvalues of the $\binom{k}{m}$ sub matrices, C_K^T, C_K, derived from C, where K is an index set of cardinality m which restricts C to those columns indexed by K. A typical sparse measurement matrix satisfying the restricted isometry property is the random Gaussian matrix $C \in R^{D \times k}$ where $r^{ij} \sim N(0, 1)$, as used commonly in previous works [49][50][51]. In the ADCOM algorithm, a very sparse random measurement matrix C with i.i.d (independent identical distribution) entries are considered based on the novel work of Achiloptas [52] According to his theory,

$$r_{ij} = \sqrt{s} \times \begin{cases} 1 \\ 0 \\ -1 \end{cases} \text{With probability of } [\frac{1}{2s}, 1-\frac{1}{s}, \frac{1}{2s}] \text{ respectively} \quad (2.20)$$

Achiloptas proved that this type of matrix with s = 1, 2 or 3 satisfies the Johnson-Lindenstrauss lemma. This matrix is very easy to compute, requiring only a uniform integer random generator. With s = 3, one can achieve a threefold speedup because only 1/3 of the data need to be processed (hence the name sparse random projections) and 2/3 of the computations can be avoided. The multiplication factor $\sqrt{s}$ can delay the process, hence no floating point value is needed and all computations amount to highly optimized database aggregation operations. This algorithm was first experimentally tested on image and text data as an application of random projection in 2001 [53].

The sparse random projections are sampled at a rate of 1/s (i.e., when s = 3, only one-third of the data are sampled), which is a considerably smaller data sample size. Statistical and empirical results reveal that, in certain cases, one does not have to sample one-third of the data in order to obtain good estimates. In fact, when the data are approximately normal, log D of the data can be used as in s = D/log instead of s = D/4, as used in this algorithm because of the exponential error tail bounds, which are common in normal-like distributions, such as binomial and gamma distributions. For certain tracking applications, such as facial recognition against a static background, and to improve robustness, it is recommend choosing s = $\sqrt{D}$, which could further reduce the portion of samples that need to be processed.

2.2.6 Sparse Measurement Matrix Representation

The fundamental definition of a sparse matrix is a matrix containing few non-zero elements and mostly zeroes in its rows and columns. Sparse matrices have several attractive properties, such as low computational complexity in both encoding and reconstruction, simple incremental updates to signals, and low storage requirements. The Bayesian framework, which is incorporated as a classifier in the ADCOM algorithm, generally assumes an a priori probability distribution that favors sparsity for the signal vector and uses a maximum a posteriori (MAP) estimator, which provides a point "best" estimate to incorporate the observation [54].

In order to represent the sparse matrix, consider an image I of size N × M and which is vectorized into a column vector X of size P × 1, where P = N*M, by concatenating the individual columns of I in order. The nth element of the image vector I is referred to as I (n), where n =1... P.

Let us assume that the basis $Psi = [Psi_1...Psi_z]$ provides a K-sparse representation of X:

$$X = \sum_{n=1}^{P} Theta(n) Psi_n = \sum_{l=1}^{K} Theta(nl) Psi_{nl} \qquad (2.21)$$

Where θ (n) is the coefficient of the nth basis vector ψ (ψ: P×1) and the coefficients indexed by 'nl' are the K-nonzero entries of the basis decomposition. Equation (2.9) can be more compactly summarized as follows

$$X = Psi \times theta, \qquad (2.22)$$

Where theta is a P×1 column vector with K-nonzero elements. If $\|\cdot\|_p$ is used to denote the l_p norm and if the l_0 norm simply counts the non-zero elements of theta, the image I is called as K-sparse.

The tracking task is formulated as a binary classification via a naive Bayes classifier with online update in the compressed domain. It is to be noted that this ability would not cost a significant decrease in tracking performance of the ADCOM algorithm.

2.2.7   ADCOM Tracking Algorithm

Before discussing the process of classification and tracking, the method is first adapted to initially select relevant features for tracking based on dimensionality reduction in the form of compression.

2.2.7.1 Haar-Like Features

The ADCOM algorithm makes use of the Haar-like features technique to determine significant features from the tracked object, which are compressed using the random sparse measurement matrix. The compressive feature extraction method with dimensionality reduction preserves the structure of the object. These features make use of the change in the contrast values between a set of adjacent pixels encrypted inside rectangular windows [55]. The intensity values of each pixel are not considered in this case. The changes in contrast between the rectangular pixel groups are used to distinguish relatively light and dark areas. Hence, two or three adjacent groups with a relative contrast variance form Haar-like features.

The concept of the use of an integral image in the form of summed area tables to improve the speed of the feature classification is incorporated in the feature extraction process. The integral image is the intermediate representation of an image frame in which simple rectangular adjacent pixel features are calculated. The integral image uses a series of these features, which are simple rectangular areas with relatively dark and light regions, to find corresponding matching patterns in adjacent groups of pixels. The integral image consists of a 2D or 3D matrix of the same size as the original image, in the form of look up tables.

Each element of the matrix represents the sum of the all the pixel intensity values located directly to the upper-left of a pixel location [56]. The integral image can be represented as follows:

$$I(x, y) = i(x, y) + I(x-1, y) + I(x, y-1) - I(x-1, y-1) \qquad (2.23)$$

Where $i(x, y)$ corresponds to the intensity of the pixel under consideration and the other terms correspond to the intensity values of the adjacent pixels in the rectangular area.

The changes in contrast between two or more rectangular groups of pixels contribute to the feature classification, which is then compressed for further processing.

2.2.7.2 Dimensionality Reduction (Lossless Compression)

Compared to traditional dimensionality reduction methods, the proposed algorithm makes no use of data-dependent parameters, nor does it require additional computation for the eigenvalue decomposition. Mathematically, the algorithm makes use of the set of input samples of features, which are sampled as positive and negative samples depending on their distance from the object tracked. These are taken into the compression process as a library of positive, negative and zero entities [57]. The algorithm then selects the features that minimize the residual output error iteratively, thus the resulting features have a direct correspondence to the performance requirements of the given task.

Furthermore, the proposed algorithm makes use of a sparse measurement matrix with which the higher-dimensional multi-feature vector spaces are concatenated to produce a lower-dimensional compressed feature set that is classified using a naïve Bayes classifier.

An example is considered to understand the dimensionality reduction and the compression process:

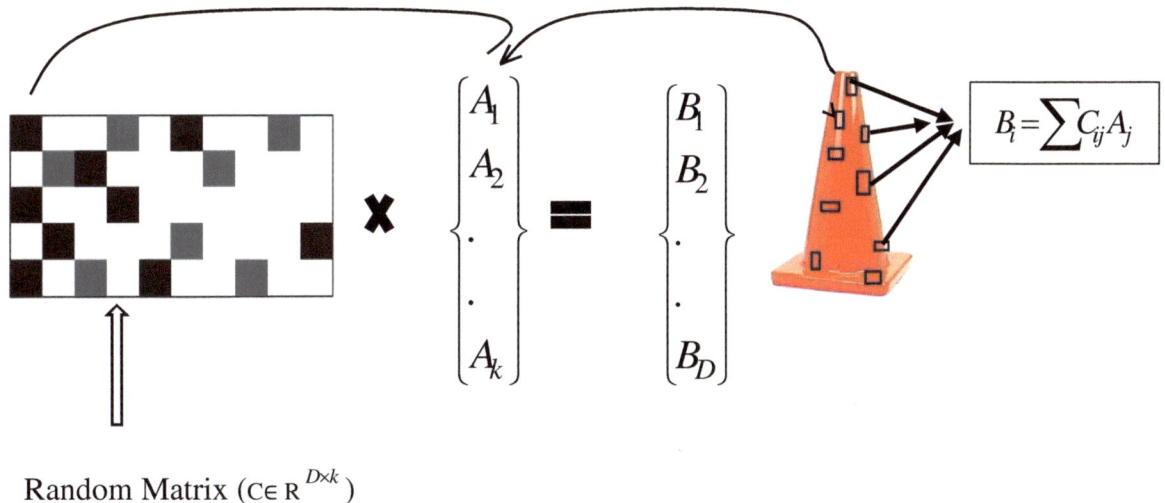

Figure 2.4 ADCOM implementation

The random matrix C consisting of gray, black and white areas indicate positive, negative and zero entries respectively. Here, consider a sample $Q \in R^{w \times h}$, which is one entry from the random matrix. Here 'w' and 'h' represent the width and height of the feature window The sample S is convolved with a set of rectangular filters at multiple scales which can defined as:

$$h_{i,j}(x, y) = \begin{cases} 1, 1 \leq x \leq i, 1 \leq y \leq j \\ 0, otherwise \end{cases}$$

Here 'i' and 'j' represent the width and height of a rectangular filter, respectively. The arrows illustrate that one of nonzero(gray or black) entries of one row of C sensing an element in A is equivalent to a rectangle filter of a particular width and height representing the pixel intensity at a fixed position of an input image. Here, the set of filtered image frames would belong to a column vector which can be concatenated as a very high-dimensional multi-scale image feature column vector A= $(A_1, A_2, ... A_k)^T$. Note that $k = (w*h)^2$ here and 'k' represents a very high dimensionality. The sparse random matrix is then used to project the higher dimensional vector to a much lower dimensional space without compromising the structure of the image data that are compressed.

The lower dimensional image feature vector B is comprised of elements that are a linear combination of spatially distributed rectangle features at multiple scales. The compressed features are computed based on relative intensity differences, which is based on a method parallel to the Haar-like feature detection. As discussed earlier in section 2.2.8.1, these Haar-like features preserve the information of the original image in a compressed form. Features obtained after compression are employed as classifiers for object tracking.

2.2.7.3   Naïve Bayes Classifier

A Naive Bayes classifier is a simple probabilistic classifier based on applying Bayes' theorem (from Bayesian statistics) with strong (naive) independence assumptions. A more descriptive term for the underlying probability model would be "independent feature model."

The Bayes theorem describes how the conditional probability of each of a set of possible causes for a given observed outcome or consequence can be computed from knowledge of the probability of each cause and the conditional probability of the outcome for each cause [58].

In simple terms, a naive Bayes classifier [59] assumes that the presence (or absence) of a particular feature of a class is unrelated to the presence (or absence) of any other feature of the same class. Depending on the precise nature of the probability model, naive Bayes classifiers can be trained very efficiently in a supervised learning setting such as in a tracking algorithm where it is used as a classifier to train the compressed vector space in order to separate the positive foreground samples from the negative background samples.

The discrete data needed for the classifier are based out of the positive and negative samples extracted by the Haar-like feature extraction. Here the input library is comprised of these extracted samples, which are distributed over a random matrix as non-zero and zero entries. The output labels are given as y= {0, 1} and a training set of i.i.d entries is obtained as $h_{i,j}(x, y)$ concatenating the input library with output labels. The naïve bayes classifier uses the training set to calculate the estimates of the probabilities $p(B_i | y)$ and $p(y)$. The estimates of the probabilities are calculated in which the number of occurrences of an event in the training set is accounted for. The bayes classifier is modelled as below:

$$H(B) = \log(\frac{P\prod_{i=1}^{n} p(B_i | y=1) p(y=1)}{P\prod_{i=1}^{n} p(B_i | y=0) p(y=0)}) = \sum_{i=1}^{n} \log(\frac{p(B_i | y=1)}{P(B_i | y=0)}) \qquad (2.24)$$

Based on the work of Diaconis and Freedman [60], most of the random projections of high dimensional random matrices are Gaussian. The mainstay of data analysis, as in this case, makes use of the lower dimensional projections to study the higher dimensional data sets. The two or higher order dimensional data sets can be represented as one dimensional histograms. According to their work, the standard measure of randomness is entropy.

The maximum entropy occurs during the Gaussian distribution for a fixed scale. According to the above theory, the probability estimates $p(B_i | y=1)$ and $p(B_i | y=0)$ are Gaussian distributed with four parameters corresponding to the mean of positive and negative features, standard deviation of positive and negative features m_1, m_0, sig_1, sig_0 respectively. According to the maximum likelihood estimation (a totally analytic maximization procedure), a method for estimating the parameters of the Bayes classifier model, the following equation set is derived.

$$m_i^1 \leftarrow lm_i^1 + (1-l)m^1$$
$$sig_i^1 \leftarrow \sqrt{l(sig_i^1)^2 + (1-l)(sig^1)^2 + l(1-l)(m_i^1 - m^1)^2} \quad (2.25)$$
$$sig^1 = \sqrt{\frac{1}{n}\sum_{k=0|y=1}^{n-1}(B_i(k) - m^1)^2} \; ; m^1 = \frac{1}{n}\sum_{k=0|y=1}^{n-1} B_i(k)$$

Some generative and discriminative algorithms make use of the training samples cropped from previous frames, which are stored and updated, but the ADCOM algorithm uses a data-independent random measurement matrix. Second, the algorithm extracts a linear combination of generalized Haar-like features in contrast to other trackers that use the whole template of the landmark for tracking purposes. These two properties of the ADCOM algorithm clearly differentiate it from other tracking algorithms.

2.3 Secondary Algorithms to support data-dependent and independent algorithms

2.3.1 Template Matching

Template matching is a simple process in which a template is matched to an image where the template is a sub-image that contains the shape or contour of the object to be tracked. The generic algorithm centers the template on an image point and sums the total number of points in the template that match those in the image. The procedure is repeated for the entire image, and the point that leads to the best match, corresponding to the maximum count, is defined as the point where the shape (given by the template) lies within the original image.

Template matching may be used when the standard deviation of the template image compared to the source image is small. Templates are most often used to identify printed characters, numbers, and other small, simple objects. Template matching is performed on either bi-level images (black and white) or grey level images depending on the application. In Simulink models, the template can be loaded from the main image during the object tracking process. The best match found in the search over the image continues the tracking of that particular portion of the original image.

Formally, template matching can be defined as a method of parameter estimation. The parameters define the position of the template. Template matching uses a specific similarity criterion for locating an object, using the correlation principle to incorporate the following equation,

$$p = \frac{\sum_x \sum_y (A_x - \overline{A})(B_{xy} - \overline{B})}{\sqrt{\left(\sum_x \sum_y (A_{xy} - \overline{A})^2\right)\left(\sum_x \sum_y (B_{xy} - \overline{B})^2\right)}} \qquad (2.26)$$

Where A and B are the 2-dimensional 'mean' of the respective image matrices, and (x,y) are the spatial coordinates within A and B. This correlation coefficient 'p', closely resembles a traditional statistical correlation, with the difference being that the traditional method is calculated in one dimension instead of two dimensions. A high correlation coefficient in a pixel-by-pixel comparison between the template and the region of interest (ROI) indicates a good match.

2.3.2   Pattern Recognition

Pattern recognition entails analyzing raw data and categorizing them into available classes or subsets. In order to perform this classification, one feature space is typically selected to represent the data in a manner that simplifies the classification task. Once the features are identified, each class or category is defined using specific models. Once data of an unlabeled object is read, the class or category of this object is determined by inferring which of the descriptions best classifies the features.

This process of detecting, describing and recognizing visual patterns has led to advances in automating several tasks like optical character recognition, scene analysis, fingerprint identification, and facial recognition. When applied to tracking hand gestures, histograms of the image measurement such as the gradient orientation or color probability of the tracked hand may be used to recognize gestures captured in the video image.

In generic applications, such as face or hand motion tracking, a system is operated in two modes. First, the system operates in a training mode in which the measurement data (e.g., gradient orientation histograms) taken from the specific tracked regions of a video frame are linked with identified gestures or motion of the object. Next, the system operates in a performance mode. In performance mode, hand or face regions of video frames are compared with the stored hand or face regions to identify the gesture or motion captured in the video frame.

2.3.3 Color Detection

Most state-of-the-art and common approaches to object detection and tracking rely heavily on intensity-based features without considering the color information in the image. This exclusion of color information is usually due to large variations in color caused by changes in illumination, compression, motion blur, and occlusions. These variations make the task of robust color description especially difficult. On the other hand, and in contrast to object detection, color has been shown to yield excellent results in combination with shape features for image classification. There are three main criteria to consider when choosing a color tracking algorithm for vision-aided navigation purposes.

Feature Combination: There exist two main approaches to combining shape and color information: early and late fusion. Early fusion combines shape and color at the pixel level, which are then processed together throughout the rest of the learning and classification pipelines. In late fusion, shape and color are described separately from the beginning and the exact binding between the two features is lost. Early fusion, in general, results in more discriminative features than late fusion since it preserves the spatial binding between color and shape features.

Photometric Invariance: One of the main challenges in color representation is the large variation in features caused by scene-accidental effects such as illumination changes and varying shadows. Photometric invariance theory provides guidelines on how to ensure invariance with respect to such events; however, photometric invariance comes at the cost of discriminative power. The choice of the color descriptor used should take into consideration both its photometric invariance as well as its discriminative power.

Compactness: Existing luminance-based object detection methods use complex representations. For example, the part-based method models an object as a collection of parts, where each part is represented by a number of histograms of gradient orientations over a number of cells. Each cell is represented by an N-dimensional vector. Training such a complex model, for just a single class, can require large gigabytes of memory space and a supercomputer to process the information and track the objects. When extending these cells with color information, it is therefore imperative to use a color descriptor or a color classifier as compact as possible to reduce both memory usage and total learning time. A specific color signal can be selected and separately processed, which can be used for tracking purposes based on its variation.

2.3.4 Edge Detection

Edge detection is the process of finding edges in an image in order to facilitate image segmentation, data compression, and image reconstruction. An edge may be regarded as a boundary between two dissimilar regions, points or sets of pixels in an image. In computer vision and image processing systems, in order to interpret an image, the separation of the image into object and background regions is a critical step. Segmentation partitions the image into a set of disjoint regions that are visually different, and are uniform and meaningful with respect to some characteristics or computed properties, such as grey level, intensity, texture or color to facilitate image analysis. Edge-based methods are the most commonly used techniques for performing image segmentation. The edges in an image are the significant characteristics that provide an indication of spatial frequency. Edge detection is commonly used for feature detection and feature extraction as part of object identification algorithms.

The edge detection process marks points in a digital image at which the luminous intensity changes sharply. In many image analysis processes, the edges of objects in the image are first detected. Edge representation of an image significantly reduces the amount of data to be processed, yet it retains useful information about the shapes of objects in the scene. The effectiveness of many image processing and computer vision tasks depends on the detection of meaningful edges.

Edge detection has proven to be a challenging task in low level image processing. Various approaches are available for edge detection, some of which are based on error minimization, maximizing an object function, neural networks, fuzzy logic, wavelet approaches, Bayesian approaches, morphology, and genetic algorithms.

2.3.4 Corner Detection

The point where two edges meet in an image frame is defined as a corner. Corner detection is based on detecting corner feature points with the primary goal of obtaining robust, stable and well-defined image features for object tracking and recognition of three-dimensional landmarks or objects. There is no image reconstruction available from the corner detection process.

Corner detection is frequently used in motion detection, image matching, tracking, image mosaicing, panoramic stitching, 3D modeling and object recognition. Corners in images contain a great deal useful information about the scene. Extracting corners accurately is significant to image processing, which can reduce many of the calculations. The corner detection procedure on object boundaries involves, first, segmenting a scene image into meaningful regions, and then extracting boundaries from the regions of interest.

2.3.5 KLT Tracker

The Kanade-Lucas-Tomasi (KLT) tracker has been studied extensively and used for feature point tracking. The KLT tracking algorithm is based on the assumption of small frame-to-frame motion of features, and initially KLT computes the optical flow of feature points and performs nonlinear optimization.

Feature tracking is a first step in many machine vision applications including optical flow, object tracking, 3D reconstruction and collision avoidance. Higher-level computer vision algorithms require, therefore, robust tracking performance without accounting for the motion of the camera. KLT feature point tracking uses template image alignment techniques. The fundamental assumption of KLT feature tracking is small spatial and temporal changes in appearance across consecutive images. Mathematically speaking, the KLT tracker solves a nonlinear optimization problem and has a limited convergence region for a true global minimum. Incorrect solutions based on local minima can be minimized when a new search region is given from an external inertial sensor instead of simply using the last tracking state.

The flow of the ADCOM algorithm or the process can be pictorially represented as shown in Figure 2.5.

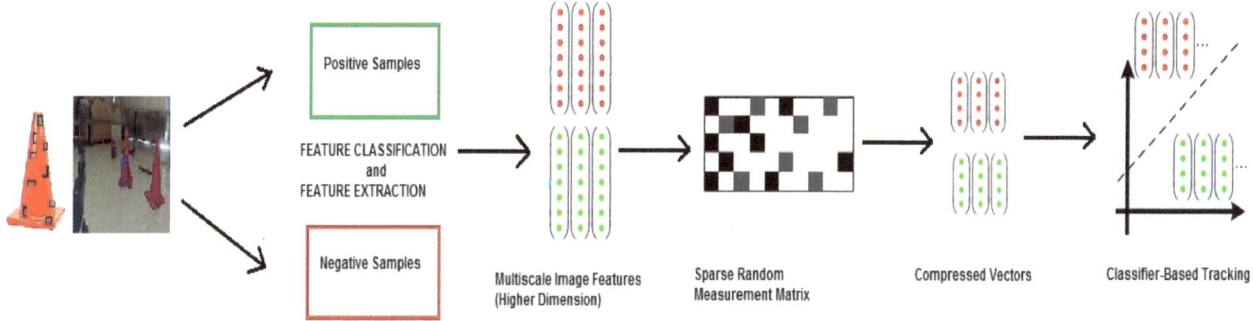

Figure 2.5 ADCOM Implementation

In brief, the algorithm observes the search window and as the initialization process occurs, it begins to recognize and identify the landmark. Once the Haar-Feature detector is employed the positive and negative samples are classified, extracted and placed in a higher multi-dimensional vector space. The compression process which works on the principle of random projection compresses(lossless compression) to a lower dimension using the random sparse matrix as descried in Chapter 2.Finally the classifier uses the compressed data to track the features in every frame as the classifier gets updated at every time step.

3 IMAGE PROCESSING

Image processing is a technique of continuously and rigorously bypassing the camera data in terms of image frames through an analysis phase thereby obtaining a processed image at the output. The analysis phase may include image display & printing, image manipulation, features extraction, image enhancement, image compression etc. An image is an array or a matrix of square pixels (discretized points in a graphical image)

As discussed in Chapter 2, the two essential image processing algorithms in consideration are CAMshift(Continuously adaptive Mean shift) and ADCOM (advanced compressive tracking).Each of the algorithms were implemented in a series of steps in order to obtain robust tracking results which pave way for further analysis. The analysis of both these algorithms are based out of experimentation and testing of four different examples. These are studied in an elaborate manner.

3.1 The 8-step DIY CAMshift remedy

The CAMshift algorithm is robust and depends entirely on the color probability or the texture of the object being tracked. It is adaptive in that it continuously changes the window size based on changes in the object properties. A trademark feature of this algorithm is that it provides robust tracking in dynamically changing environments. The implementation of the algorithm is carried out using the following sequence, executed in MATLAB.

Step 1: Initialize the directory in which the camera data is stored as N frames at a certain sample rate (frames per second, fps) along with the initial threshold level and pixel rate of expansion for the search window.

Step 2: The search window size and location are defined carefully after close inspection of the position of the target/landmark in the camera frame.

Step 3: The Hue (color channel) is extracted as a 1D histogram for processing in the region of interest as it highly suitable for generating a color probability distribution.

Step 4: Initialization and calculation of the zeroth, first and second order moments along with the calculation of the centroid of the tracked object based on these moments.

Step 5: The covariance matrix is defined using the single value decomposition method to estimate the width, height and orientation of the tracked object.

Step 6: Calculate the orientation along the longitudinal and lateral axes. Steps 5 and 6 are non-trivial as they can be used to study the change in the properties of the object and how the algorithm adapts to these substantial changes.

Step 7: The new window size is calculated after the convergent values (based on the centroid) are computed. This calculation is based on the area of the probability distribution making use of a scaling factor. An appropriate scaling factor is one which does not generate unreasonably large increases or decreases in the size of the search window.

Step 8: Display and print the tracking and scaling results, store the tracked frames and plot a map based on the tracked centroids.

These examples demonstrate the centroid tracking ability of the CAMshift algorithm. In each example, it can be noted that the background is dynamically changing and is not a static background. All the experiments conducted included a static object on the foreground but a dynamic background and a moving camera either mounted on a vehicle or manually held.

The red ball was a unique landmark or target when compared to the background in which it was tracked. The reasoning behind this claim is that the color probability histogram was able to obtain a dense probability function as the red color was easily distinguishable.

Specifically, the cone tracking experiments were conducted under varied conditions with numerous constraints in the environment. Some of the constraints involved the varying light intensity on the target from the environment, similar sized cones in the scene with the same color probability, and multiple landmarks in the vicinity of the original cone, and camera motion or vibration due to mounting on the quad copter.

The final set of experiments was conducted with a rock as a static foreground landmark using a camera mounted on an unmanned ground vehicle which runs on a high-speed dual core processor. The controller for the ground vehicle was already programmed and operated through a separate keyboard that provides commands for maneuvering the vehicle. The camera was internally programmed for 15 fps, although accurate centroid tracking can be achieved at lower frame rates. The downside to this approach lies in the fact that if there is constant change in the illumination, motion changes, or occlusions, a robust estimation of the centroid is not possible.

3.2 The 5-step DIY ADCOM solution

As discussed in Chapter 2, the advanced compressive (ADCOM) tracking algorithm continuously makes use of the features derived from the landmark or target, classifying them as positive or negative features. A search window established around this landmark facilitates the continuous extraction of positive and negative features. The initial determination of the search window size, which is specified in pixels, is an important factor affecting the performance of the algorithm. A poorly sized search window can extract features from different objects if the background is almost the same throughout. Smaller search window sizes help in the extraction of these features as it would be concentrated to a localized area around the target.

The steps involved in the ADCOM tracker are as follows:

Step 1: Load the frame data, initialize the search window, and input the corresponding parameters for the negative and positive samples (i.e., the number of samples and radial scope).

Step 2: Determine the classifier parameters, the rectangle window parameters, learning rate, and positive/negative classifier parameters.

Step 3: Using a Haar-like feature detector, extract the features from the landmark and calculate a sample template.

Step 4: Extract the positive and negative features and extract the data from them for updating the classifier.

Step 5: Display the tracking results and update the classifier for the next consecutive frames.

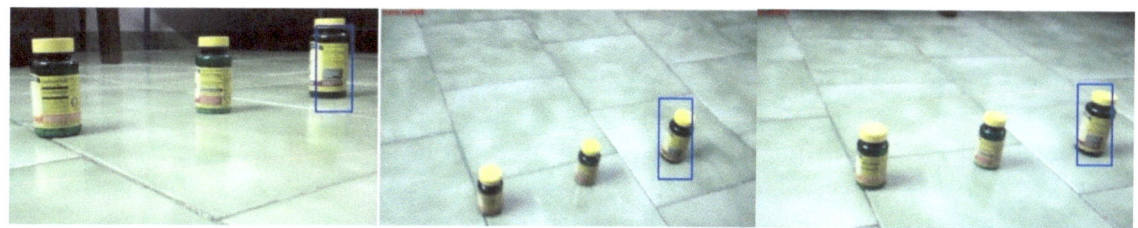

It was observed from the experiment that the tracker separated the positive and negative samples inside the search window and hence any change in the orientation of the camera with respect to the landmark (red ball) did not induce drift while tracking. The tracking ability did not get affected by the introduction of a flesh-toned hand as it was tracking the positive samples from the red ball, which was constantly updated using the Bayes classifier.

The key difference in performance lies in the fact that the ADCOM algorithm incorporates a structured classifier method that is partially dependent on the color intensity of the landmark, unlike the CAMshift algorithm, which is purely driven by the color probability density function. The ADCOM algorithm is more adaptable than the CAMshift algorithm for applications in which there are multiple, similar moving targets and only one of these is to be tracked continuously.

The cone tracker experiment was carried out under varied illumination as the light intensity in the environment continuously changed owing to various factors such as the intrusion of the sun's rays and temporary yellow light set up in the environment. It is to be noted that there was a large set of templates that were extracted as the number of unnecessary targets in the background image was high in number in comparison to the other experiments. The tracking shows that only the selected landmark was successfully tracked until a point where the target was completely lost in the frame due to the sudden change in the attitude of the aerial vehicle carrying the camera.

The tracking capability of this algorithm was tested to a certain extreme while conducting this particular experiment because three similar landmarks were in the scene with the objective of tracking only one landmark/target. The prominent difference between the CAMshift and the ADCOM algorithms was clearly shown in this particular test as the ADCOM algorithm was able to track over the entire library of frames in spite of introducing similar targets in the environment.

To summarize the aforementioned algorithms, the table below vividly shows their core functionalities.

CAMshift Algorithm	ADCOM Algorithm
Data Dependent Algorithm	Data Independent Algorithm
Centroid-based Tracking with Higher Computational Time	Feature-based Tracking with Lesser Computational Time
Based out of Mean-shift Theory	Based on Johnson-Lindenstrauss Lemma principle for Random Projection
Search window is Adaptive in nature	Non-adaptive nature of Search window
Determination of a Color Probability Distribution [Histogram Back-Projection/Look up Table]	Feature Classification, Feature Extraction, Dimensionality Reduction and Classifier-based Feature Tracking [Summed Area Table]
Kernel estimation as a classifier(color PDF)	Naïve Bayes classifier
Scaling and Orientation Calculation	No such provision or requirement [12]

4 NAVIGATION FILTERS

Navigation is defined as the process of determining the current motion parameters of a vehicle, such as acceleration, velocity and position of the center of mass and planning solutions to traverse a path or a map based on the estimated parameters. The system that provides and supports this process is termed a navigation system. One of the most commonly used navigation systems, which has been developed for a wide range of vehicles, is the Inertial Navigation System (INS) [61]. This system integrates the accelerations and angular rates provided by an Inertial Measurement Unit (IMU) to compute the position, velocity, and attitude of the vehicle. Because IMU solutions for position and attitude are subject to drift errors that build over time, Global Positioning System (GPS) data is typically fused with IMU data in a Kalman filter within the INS in order to provide optimal estimates of the vehicle position, velocity and attitude.

4.1 Kalman Filter

The Kalman filter is an optimal recursive estimator in which the estimated state from the previous time step and the current measurement are required to compute an estimate of the current state. The filter is an implementation of the least squares algorithm and infers the parameters or states of interest from uncertain observations. The recursive nature of the filter [62] [63] allows it to process the data sequentially as they arrive, providing real-time estimation. The process of finding an optimal estimate from noisy data requires filtering the noise, which is accomplished in the Kalman filter assuming Gaussian white process and measurement noise. The Kalman filter not only filters the data measurements, but also projects these measurements onto the state estimate. The Kalman filter is a subset of the Bayes filter where a linear relationship is established between the assumption of the Gaussian noise distribution and the current state to the previous state imposed.

A set of mathematical equations are used to implement the prediction and the measurement update phases in such a manner that the filter minimizes the estimated error covariance when the condition of the linear-Gaussian dependency is met. The Kalman filter has the ability to combine the subsystems based on knowledge of the sensor measurement noise covariance and the process noise covariance.

The Kalman Filter provides a solution to the linear-quadratic problem [64], which is the problem of estimating the current or present state of a linear dynamic system perturbed by white Gaussian noise. This is accomplished by using measurements that are linearly dependent on the states of the system but are also subject to Gaussian noise. The resulting predictor-update estimator is a recursive optimal solution with respect to any quadratic function of estimation error based on the error covariance matrix[65][66].

A simplified equation to represent the Kalman filter is given below, which shows that the Kalman filter finds the optimal averaging factor for each consecutive state, and the filter utilizes previous state estimates states to compute the current state estimate[67].

$$\bar{X}_k = K_k . Z_k + (I - K_k) . \bar{X}_{k-1} \qquad (4.1)$$

The fundamental purpose is to find the estimate of a state vector X. The state estimate at the current time step is given by $\bar{X}_k$ where Z_k and K_k represent the measurement and Kalman gain respectively and $\bar{X}_{k-1}$ is the previous state estimate [68]. A simplified step-by-step guide to the entire Kalman filter process is given below.

The first step involves building a system model, which is used to propagate the state from one time step to the next. The model includes the state dynamic equation and a measurement equation that expresses the measurement as a linear function of the state:

$$x_k = Ax_{k-1} + Bu_k + w_{k-1} \qquad (4.2)$$

$$z_k = Hx_k + v_k \qquad (4.3)$$

Equation (4.2) expresses X_k as a linear combination of its previous value plus a control signal u_k subject to additive process noise. In Equation (4.3), the measurement is expressed a linear function of the state subject to additive measurement noise. The process and measurement (or sensor) noise are assumed to be Gaussian and statistically independent. The filter requires estimates of the mean and standard deviation of the noise functions, W_{k-1} and V_k, which represent the process and measurement noise respectively. The Kalman filter is frequently applied with success in cases where the assumptions of independent, Gaussian process and measurement noise do not strictly hold. An initial value for the state estimate must also be provided. The Kalman filter process is then implemented in two stages: the prediction or time update phase and the measurement update or correction phase. At every time step, both stages are implemented to estimate the next state and update the filter parameters.

The first step involves projecting both the most recent state estimate and an estimate of the error covariance (from the previous time period) forward in time to compute a predicted (or a priori) estimate of the state at the current time. The second step involves correcting the predicted state estimate that was calculated in the first step by incorporating the most recent process measurement to generate an updated (or a posteriori) state estimate. The Q and R matrices represent the covariance of the process and output noise, which are specified when initializing the filter. This process is iterated for every time step. The two stage Kalman filter process is summarized in Figure 4.1

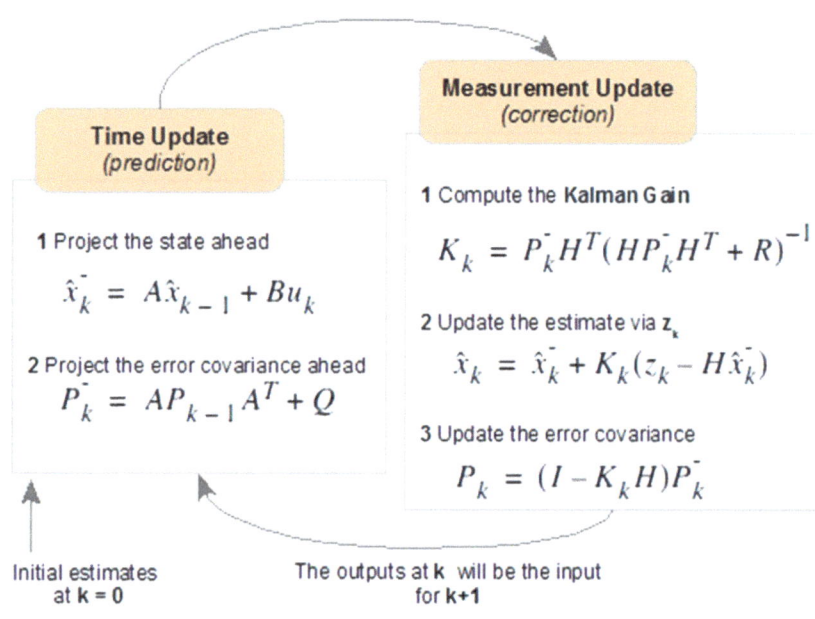

Figure 4.1 Kalman Filter Process.

4.2 Extended Kalman Filter

The Extended Kalman filter (EKF) provides an approximation of the optimal state estimate. The nonlinear system dynamics are approximated by a linearized version of the nonlinear system model around the previous state estimate. For this approximation to be valid, the linearization should be a good approximation of the nonlinear model in the vicinity of the state estimate. The EKF is not exactly an optimal filter but rather is implemented based on a set of approximations.

A non-linear discrete time process with input and measurement noise model is shown in figure 4.2 below.

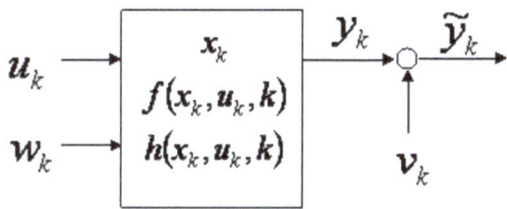

Figure 4.2 Non-linear to Linear Model

The non-linear representation can be written in the standard state space form as,

$$x_k = f(x_{k-1}, u_k, k) + w_{k-1} \qquad (4.4)$$

$$y_k = h(x_k, u_k, k) \qquad (4.5)$$

$$\tilde{y}_k = y_k + v_k \qquad (4.6)$$

In these equations,

1. 'k' denotes the current discrete time step (with 'k-1' representing the previous time step)

2. u_k is a vector of inputs, x_k is the state vector, which may be observable but not measured, y_k is a vector of the modeled process outputs and $\tilde{y}_k$ is a vector of the measured process outputs.

3. w_k and v_k are the process and measurement noise respectively. They are assumed to be zero mean Gaussian noise with covariance Q_k and R_k respectively.

4. f(.) and h(.) are generic nonlinear functions relating the past state, current input, and current time to the next state and current output respectively.

Given the inputs, measured outputs and necessary assumptions on the process and output noise, the purpose of the extended Kalman Filter is to estimate unmeasured states (assuming they are observable) and the actual process outputs. Observability is the measure of how well the estimated unmeasured internal states in a system can be determined based on the knowledge of the external outputs or the system response.

Similar to the Kalman filter, the extended Kalman Filter uses a 2-step predictor-corrector algorithm. The first step involves projecting both the most recent state estimate and an estimate of the error covariance (from the previous time step) in order to compute a predicted (or a priori) estimate of the states at the current time. The second step involves correcting or updating the predicted state estimate calculated in the first step by incorporating the most recent process measurement to generate an updated (or aposteriori) state estimate. However, due to the nonlinear nature of the process being estimated the covariance prediction and update equations cannot directly utilize the nonlinear functions f(.) and h (.). In order to solve this problem, Jacobian matrices are defined with respective to the states.

$$F_k = \frac{\partial f}{\partial x}\bigg|(\hat{x}_k, u_k, k)$$

$$H_k = \frac{\partial H}{\partial x}\bigg|(\hat{x}_k, u_k, k)$$

The EKF predictor and the corrector steps are then given below:

Step 1: *PREDICTION*

$$\hat{x}_k^- = f(\hat{x}_{k-1}^-, u_k, k)$$

$$P_k^- = F_{k-1} P_{k-1} F_{k-1}^T + Q_k$$

Step 2: *CORRECTION / UPDATE*

$$K_k = P_k^- H_k^T (H_k P_k^- H_k^T + R_k)^{-1}$$

$$\hat{x}_k = \hat{x}_k^- + K_k (\tilde{y}_k - h(\hat{x}_k^-, u_k, k))$$

$$P_k = (I - K_k H_k) P_k^-$$

In these equations, P_k is an estimate of the error covariance and K_k is the Kalman gain matrix. After both the prediction and correction steps have been performed, $\hat{x}_k$ represents the current state estimate.

Both $\hat{x}_k$ and P_k are stored and used in the predictor step for the next time step. However, unlike the standard Kalman Filter, the extended Kalman Filter is not guaranteed to be optimal in any sense. Further, if the process model is inaccurate due to the use of the Jacobians, which represent a linearization of the model, the extended Kalman Filter can diverge leading to very poor estimates. In practice, when used a reasonably accurate model is developed, the extended Kalman Filter often leads to reliable state estimation.

4.2.1 Experimental EKF for Vision-Aided Navigation

A simple EKF implementation of EKF is provided and modified for the actual experimentation (simulation) phase of the project. One of the earliest applications of the EKF [68] [69] was to solve the issue of tracking flying objects from the ground.

In this example of object tracking, at each time step the tracked object or target (either dynamic or static) has a measured range and bearing from the observer. In general cases, the observer is considered to be the location of a radar device tracking the object as seen in the figure 4.3.

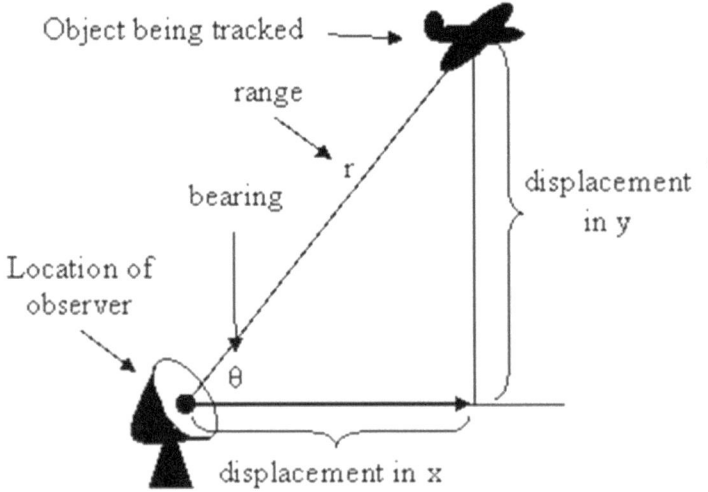

Figure 4.3 Example of tracking an aerial vehicle from the ground [71]

The range and bearing of the tracked object are represented in terms of displacements, which are given as the distances from the observer in both x and y directions. In this particular tracking problem, the state estimates are not only the x and y displacements of the target or object but also its linear velocities in x and y directions. These four states are estimated once provided with measurements of range and bearing that are subject to Gaussian white noise. The EKF is suitable for this type of problem as the displacements and linear velocities are nonlinearly related to the measured range and bearing.

For experimental purposes, a 4-state EKF can be developed to determine the tracking ability of a vehicle towards a target or a landmark at a given or determined position in the earth reference frame. This scenario is quite similar to that depicted in Figure 4.3 except the target is now fixed and the observer is now a vehicle that is in motion. The filter makes use of the four-state Equation (4.7) and body centered accelerations in the x and y directions (as measured by accelerometers), given in Equation (4.8). Equations (4.9) and (4.10) represent the linearized form of these equations to be used in the Kalman filter for state estimation. This filter employs measurements of the bearing angle and the range, which can expressed as functions of the state vector using the Cartesian to polar transformation. Equation (4.11) represents the range and bearing for a target at the location (x_T, y_T).

$$X = \{x, v_x, y, v_y\}^T \tag{4.7}$$

$$U = \{a_x^B, a_y^B\}^T \tag{4.8}$$

$$\hat{X}_{k+1} = f_k(X_k, U_k) \tag{4.9}$$

Linearized form is given as:

$$\hat{X}_{k+1} = A_k X_k + B_k U_k \tag{4.10}$$

$$\begin{aligned}Range &= \sqrt{(x-x_T)^2 + (y-y_T)^2} \\ Bearing\,(in\,rad) &= \arctan\left[\frac{(y-y_T)}{(x-x_T)}\right]\end{aligned} \tag{4.11}$$

The 4 state filter can be easily be augmented into a 6-state EKF for estimating the position and velocities in 3D, which requires a modification in the form of the Cartesian to spherical transformation to include an elevation angle. Note that the Jacobian for the state and measurement variables must be calculated during the state estimation process.

A 9-state EKF can then be formulated with a state vector that is comprised of the inertial position, velocity and Euler angles. The control inputs in this formulation are the body-centered accelerations (as provided by accelerometers) and the Euler rates (as provided by rate gyros). The IMU is used to capture these inertial data of a vehicle for navigation purposes. It is to be noted that the filter incorporates image processing data as part of the estimation as the intended application is for GPS-denied locations. The position data of the object tracked in the image frame (camera frame) has to be transformed in order to determine an accurate location with respect to the vehicle body frame. The range to the target, while not directly measured using vision, can be inferred from a priori knowledge of the size of the target and the number of pixels the target covers in the image frame. The following steps should be carried out while implementing the 9 state Extended Kalman filter for the purpose of vision-aided navigation:

Step 1: Determination of the vector distance of the target object from the camera mounted on the vehicle and the vector distance of the camera and the center of gravity/origin of the vehicle.

These vectors are represented as $\vec{r}_T^E = [x_T, y_T, z_T]^T$ and $\vec{r}_{C/B}^B = [x_{C/B}, y_{C/B}, z_{C/B}]^T$ respectively. The inertial to body frame direction cosine rotation matrix (R_E^B) is estimated in the process, which is a function of the Euler angles. The body to camera rotation matrix R_B^C is fixed and known based on the camera mounting onto the body of the vehicle.

Step 2: The states, control inputs and the estimated inertial data are defined as

$$(\text{Inertial Position}) \begin{cases} X_{k+1} = X_k + V_{xk} \partial t \\ Y_{k+1} = Y_k + V_{yk} \partial t \\ Z_{k+1} = Z_k + V_{zk} \partial t \end{cases}$$

$$(\text{Inertial Velocity}) \begin{cases} V_{xk+1} = V_{xk} + a_x^E \partial t \\ V_{yk+1} = V_{yk} + a_y^E \partial t \\ V_{zk+1} = V_{zk} + a_z^E \partial t \end{cases}$$

$$(\text{Euler Angles}) \begin{cases} Phi_{k+1} = Phi_k + (p + q \sin Phi_k \tan Theta_k + r \cos Phi_k \tan Theta_k) dt \\ Theta_{k+1} = Theta_k + (q \cos Phi_k - r \sin Phi_k) dt \\ Psi_{k+1} = Psi_k + [(q \sin Phi_k + r \cos Phi_k) \sec Theta_k] dt \end{cases}$$

$$\begin{Bmatrix} a_x^E \\ a_y^E \\ a_{z+g}^E \end{Bmatrix} = R_B^E \begin{Bmatrix} a_x^B \\ a_y^B \\ a_z^B \end{Bmatrix}$$

$\uparrow$ $\qquad$ $\uparrow$
(Inertial $\quad$ (IMU data)
Accel.)

$$\text{where } R_B^E = \begin{bmatrix} \cos Psi & -\sin Psi & 0 \\ \sin Psi & \cos Psi & 0 \\ 0 & 0 & 1 \end{bmatrix} \begin{bmatrix} \cos Theta & 0 & \sin Theta \\ 0 & 1 & 0 \\ -\sin Theta & 0 & \cos Theta \end{bmatrix} \begin{bmatrix} 1 & 0 & 0 \\ 0 & \cos Phi & -\sin Phi \\ 0 & \sin Phi & \cos Phi \end{bmatrix}$$

Step 3: The extended Kalman filter process is then implemented. The covariance of the process and output Gaussian noise are determined as well as the Jacobian matrices of all the input states and process outputs. This is done in order to establish a linearized state space equation as mentioned above. The target position estimated from the image processing algorithm is correlated with the vector distance or range between the camera and the target given by $\vec{r}_T^{\,c} = [x_{T/C}, y_{T/C}, z_{T/C}]^T$.

Step 4: The prediction and correction phases are carried out as discussed above based on the Ricatti's equation.

The measurement equations can be observed as follows:

$$Range = \sqrt{(x-x_T)^2 + (y-y_T)^2 + (z-z_T)^2}$$

$$Azimuth(in\ rad) = \arctan\left[\frac{(y-y_T)}{(x-x_T)}\right]$$

$$Zenith\ /\ Inclination\ Angle(in\ rad) = \arctan\left[\frac{\sqrt{(x-x_T)^2 + (y-y_T)^2}}{z-z_T}\right]$$

Where (x_T, y_T) represents the position of the landmark on the ground

The EKF described above is proposed to be used in the navigation system where image processing algorithms are applied for the purpose of object tracking and autonomous navigation in the absence of the Global Positioning System. This process is depicted in the Figure 4.4.

In conclusion, for the purpose of vision-aided navigation with typically nonlinear dynamic systems, where both the system dynamics and the measurements are nonlinear, state estimation is provided using a 9-state EKF implementation.

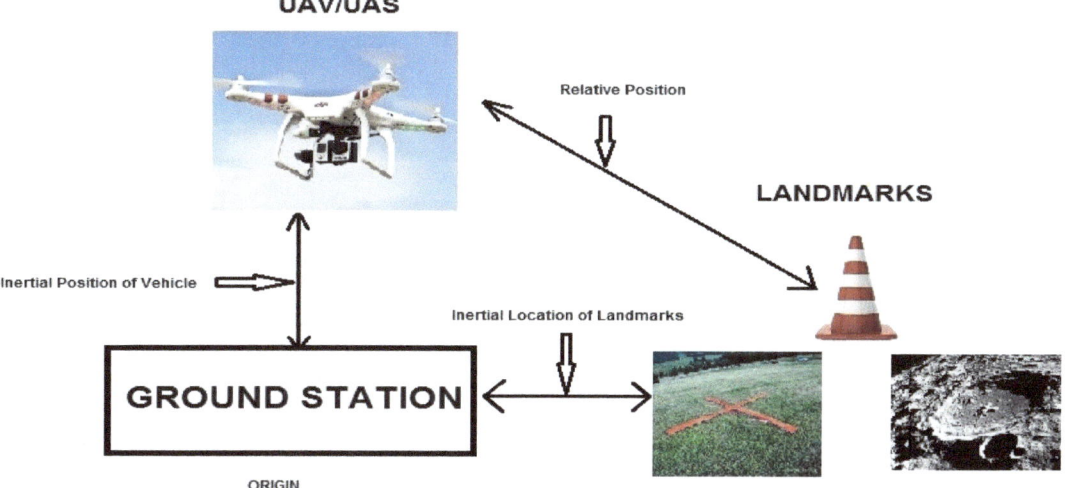

Figure 4.4 Depiction of EKF logic

The EKF has some limitations, which should be kept in mind when developing a model and performing state estimation:

1. The linear and quadratic transformations produce reliable results only when the error propagation can be well approximated by a linear or a quadratic function. If this condition is not met, the performance of the filter can be extremely poor. In the worst case scenario, the state estimates can diverge altogether.

2. The Jacobian matrices need to exist so that the transformation can be applied. There are cases for which these Jacobians cannot be defined. For example, a system that is jump-linear has discontinuous parameters.

3. In many cases, the calculation of Jacobian and Hessian matrices (in case of a second-order Kalman filter implementation) can be difficult processes subject to computational errors (model development and software implementation). These errors are often difficult to identify as it is difficult to see which parts of the system produces the errors simply by viewing the estimates, especially since the performance of the filter is uncertain.

5 SYSTEM VALIDATION

An Unmanned Ground Vehicle (Corobot Classic CL4) was used as a common platform to test the navigation filters which use the vision data. This vehicle was carrying few sensors for data acquisition and storage. Figure 5.1a and 5.1b shows the UGV used for the process of collecting data for both image processing and navigation filter design. Some of the features of the Corobot are listed below:

Basic Robot Control:

- The motor controller on-board enables to control speed of each sides of motors from -100% to 100% of total speed of motors.

- The high-speed encoder board enables to measure high-resolution ticks of encoders of front left wheel and front right wheel. If needed we also accommodate tick data of back wheels.

- The interface kit on-board allows mounting additional 5 analog sensors, and has 8 digital I/O pins. These numbers can be easily increased upon customer's request

Software:

- Easy robot control software to drive robot forward, back, left and right at different ranging speeds.

- Event driven notifications on change in encoder ticks or sensor values.

- Source code or API available in programming languages, such as, C++ and C#. For other programming languages please contact us if needed.

- Available for both the Operating Systems: Windows 7 and Ubuntu 12.04 (ROS compliant)

Figure 5.1 Corobot Classic CL4[72]

The Inertial Measurements were recorded using a MicroStrain 3dm-gx-45 module which has an attitude heading reference system in it and has an extended Kalman filter running continuously to integrate the data given by the gyros, accelerometer and magnetometer. Figure 5.2 shows the microStrain being used for experimental purposes. Also some of its significant features are listed below.

- precise position, velocity and attitude estimations
- high-speed sample rate & flexible data outputs
- high performance under vibration and high accelerations(g)
- smallest, lightest industrial GPS/INS available
- simple integration supported by SDK and comprehensive API

Figure 5.2 MicroStrain 30dm-gx3-45[73]

Vision data collection and storage:

Two different vision sensors were used for the purpose of tracking a landmark. These were either hooked onto the ground vehicle or an aerial vehicle in order to record video/image frames of the landmark in an environment.

GoPro HD Hero2

- The 11 megapixel 1/2.3-inch sensor captures video in resolutions from 720p HD (1280 x 720 pixels) at 30 or 60 frames per second to 1080p (1920 x 1089) at 30 frames per second in NTSC.
- Drop the resolution down to WVGA (848 x 480 pixels) and you can record up to 120 fps, perfect for super slow motion.
- f/2.8 fixed focus lens providing up to 170 degrees field of view

Figure 5.3 (a) GoPro Hero2 (b) Microsoft LifeCam HD-3000[74]

Microsoft Life Cam HD-3000

- A basic webcam for High definition video recording and streaming at 720p HD with 16:9 widescreen resolution
- Live video recording and easy to configure when connected to a processor

Figure 5.4 and 5.5 show the 3D path tracked by the vehicle and real-time GPS latitude-longitude data transformed in the North East Down plane. The path was verified using camera imagery as well. For the scenarios considered in theses filter experiments, the ground vehicle was driven in a 7*7 square meter area.

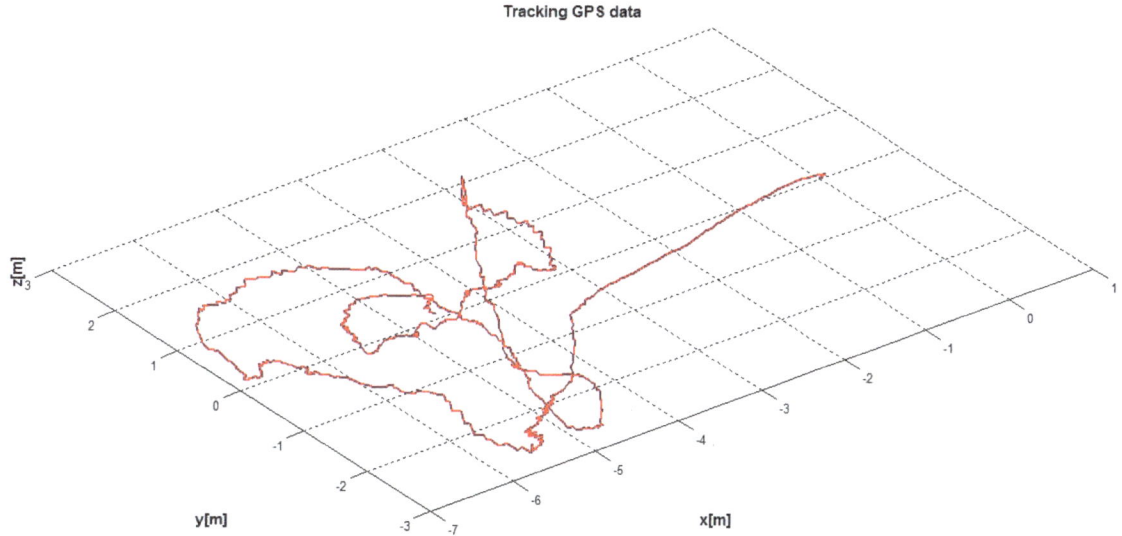

Figure 5.4 Path of the UGV

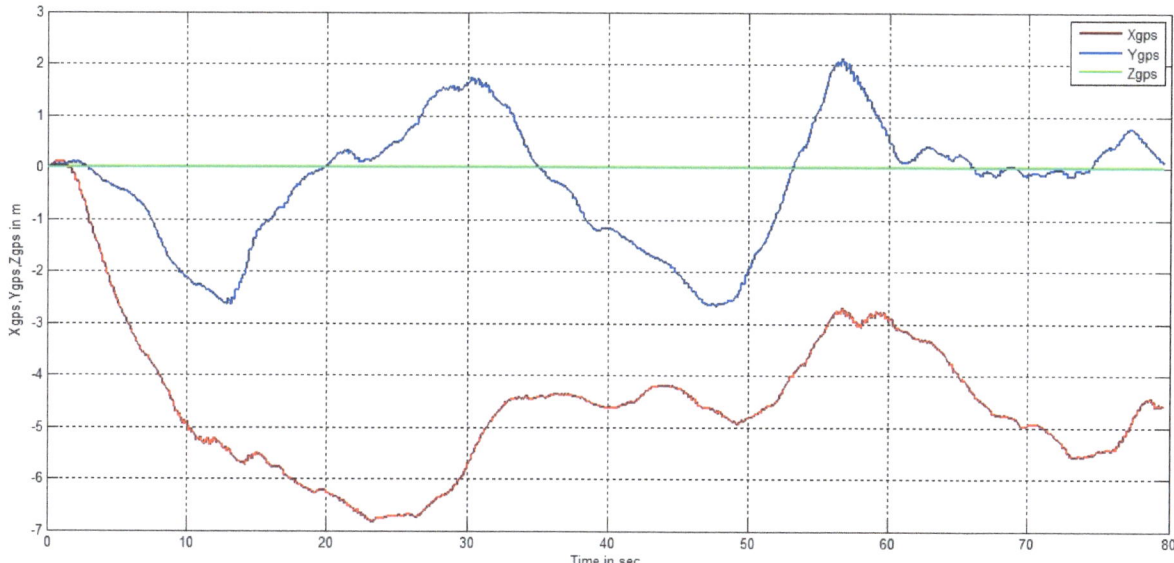

Figure 5.5 Raw GPS Lat-Long-Altitude data in NED

Figure 5.6 represents the Euler angles estimated using the angular rates measured by the gyros. Due to unleveled ground a very small roll and pitch angles were observed as the vehicle was traveling along its path.

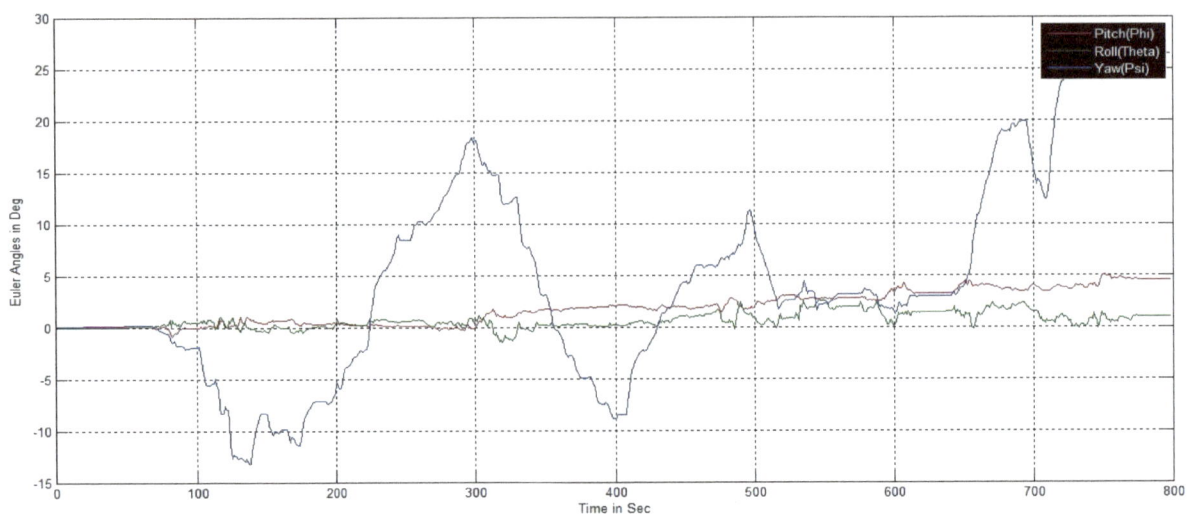

Figure 5.6 Euler Angles

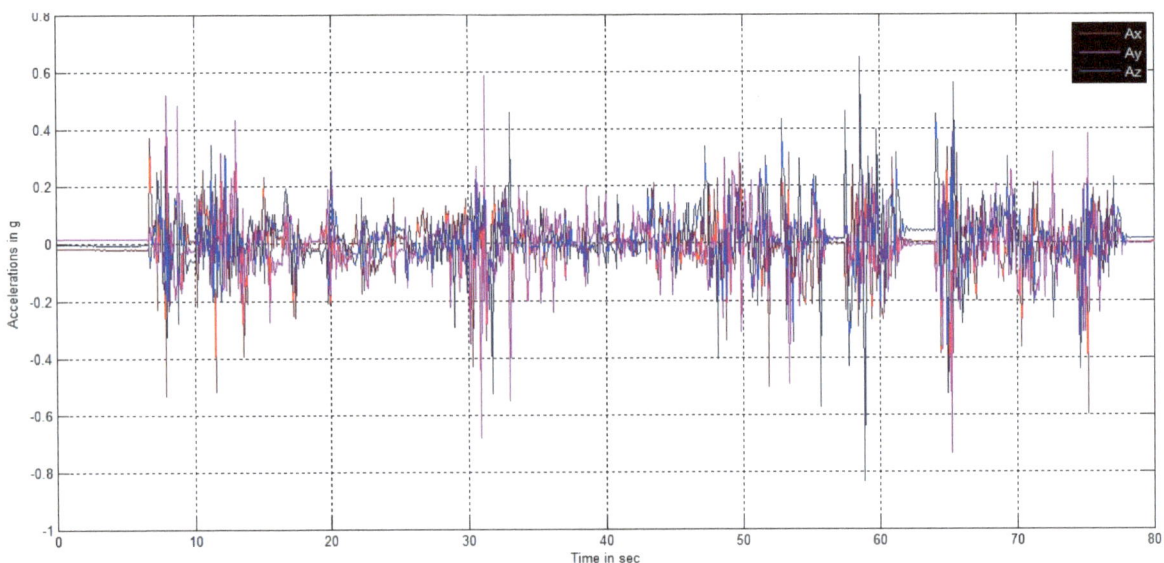

Figure 5.7 Accelerometer data from MicroStrain

Figure 5.7 represents the raw accelerometer data from the MicroStrain on the ground vehicle. The noisy data is being used for the EKF for the estimation of various states. The 'g' value is subtracted from the accelerometer data in Z axis in all the simulations.

5.1 Position And Velocity Estimates Using 4 State EKF (With Accelerometer data)

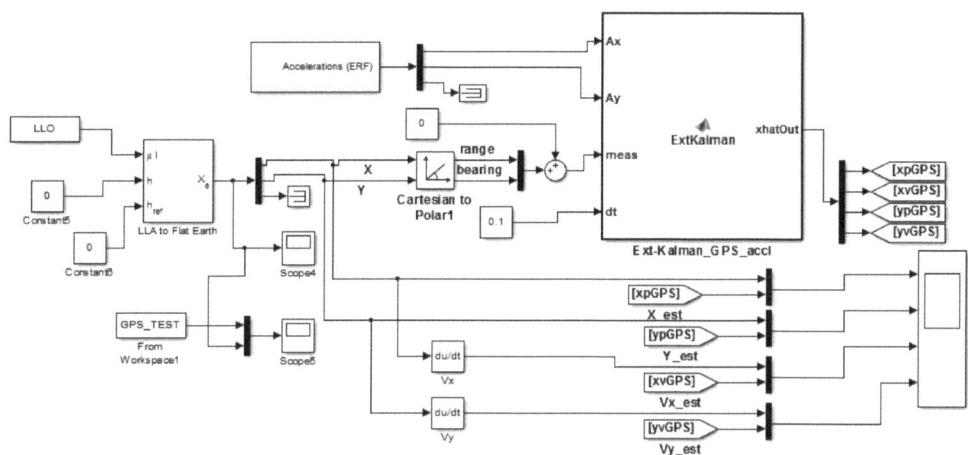

Figure 5.8 Simulink Model with Accelerometer data

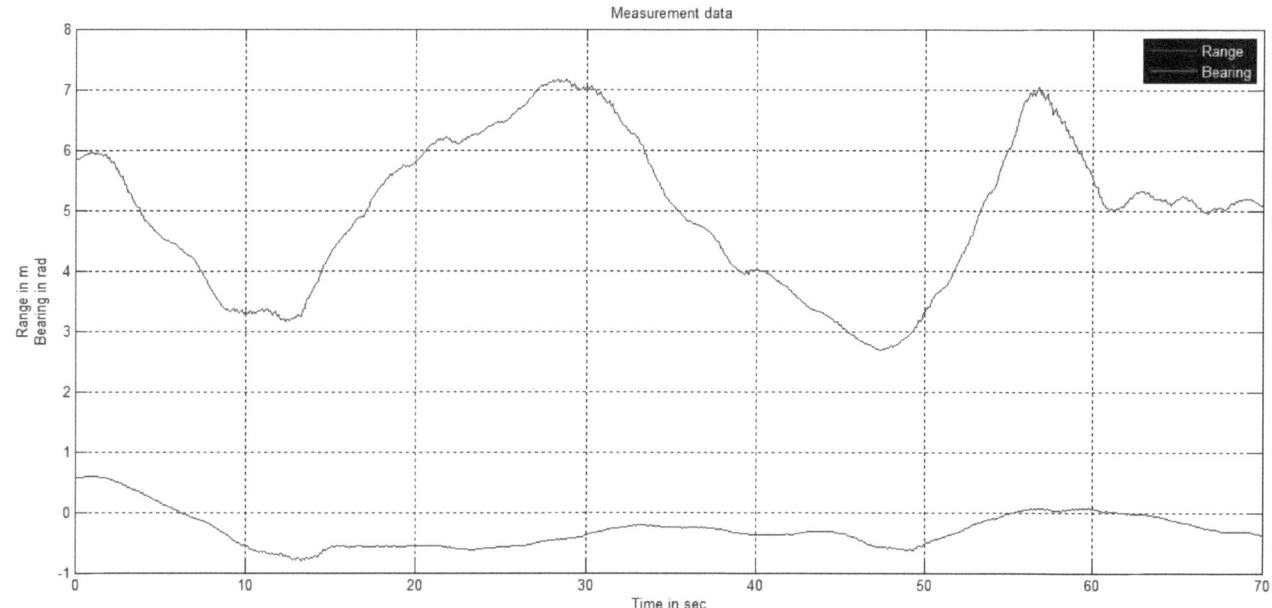

Figure 5.9 Measurement plot

Since the ground vehicle basically travels a 2D trajectory, the 4-state Kalman filter was implemented. This filter makes use of the accelerometer data for the 'U' control matrix described in equation 4.10 from chapter 3. Figure 5.8 shows the implementation of 4 state EKF including the accelerometer data.

There is a transformation made from body reference frame to earth reference frame using the Direction Cosine Matrix which is derived from the Euler angles obtained by integrating the rate gyro data. Figure 5.9 represents the measurement data used inside the filter to calculate the measurement residual and also to update the states.

In this example, target measurements, which would be obtained using image processing in practice are simulated by selecting a target location and measuring the relative target location using GPS data to represent the true position of the vehicle referred to as the ground truth.

POSITION ESTIMATES

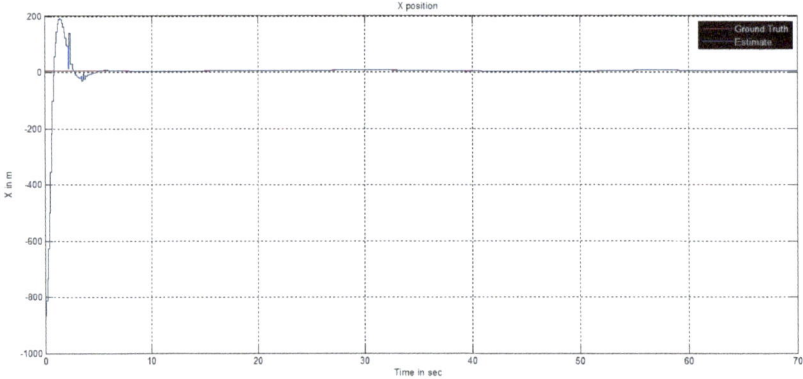

Figure 5.10 X Position

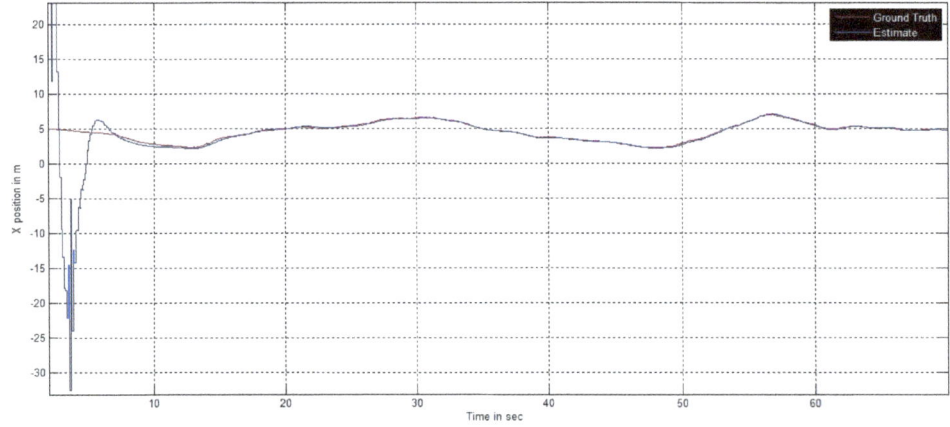

Figure 5..11 X Position zoom

Figure 5.10 and figure 5.12 represents the position estimates in the x-y direction compared to the x-y position measured by the GPS (in flat earth) which is considered as the ground truth. It can be observed later in this chapter that the performance of the filter is robust when the accelerometer data is included.

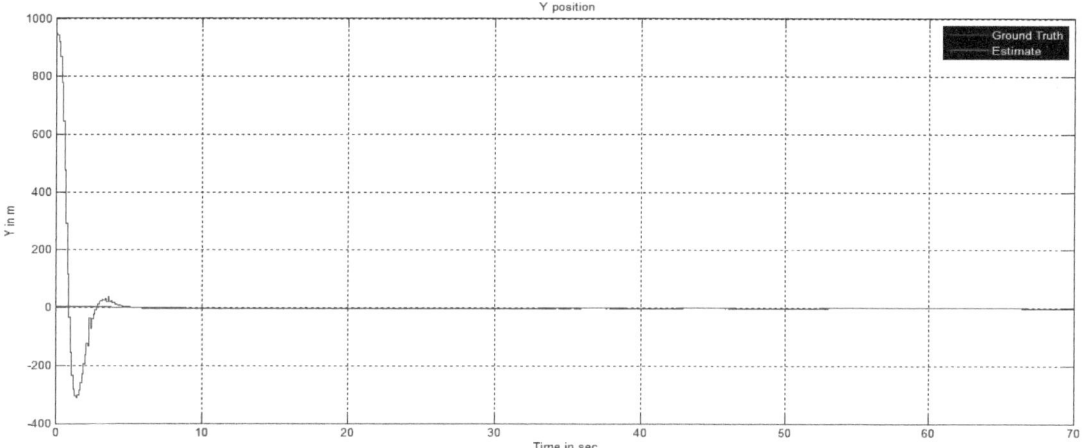

Figure 5.12 Y Position

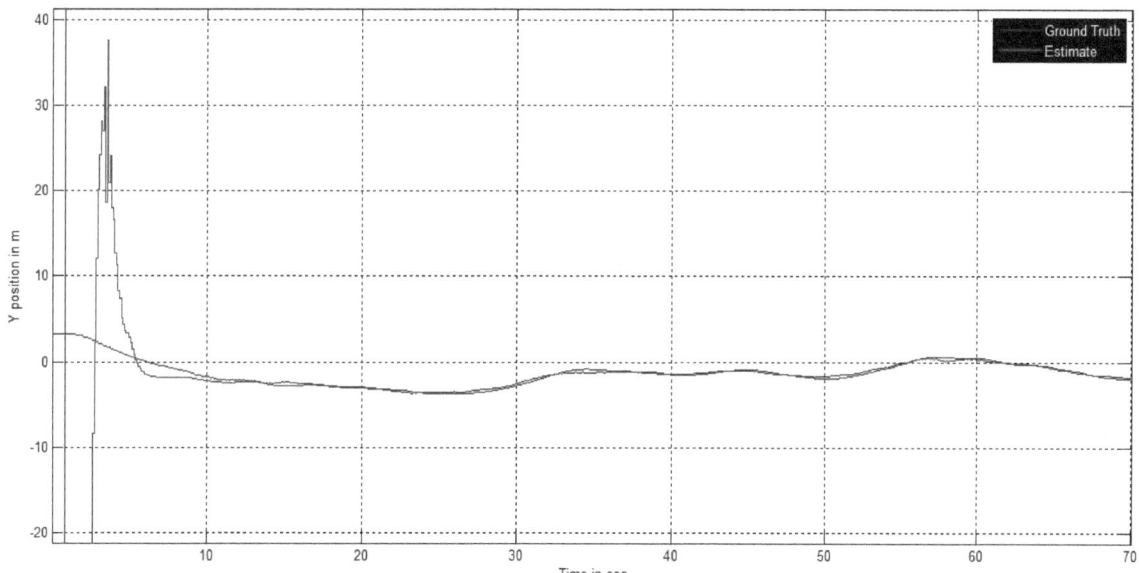

Figure 5.13 Y position zoom

VELOCITY ESTIMATES:

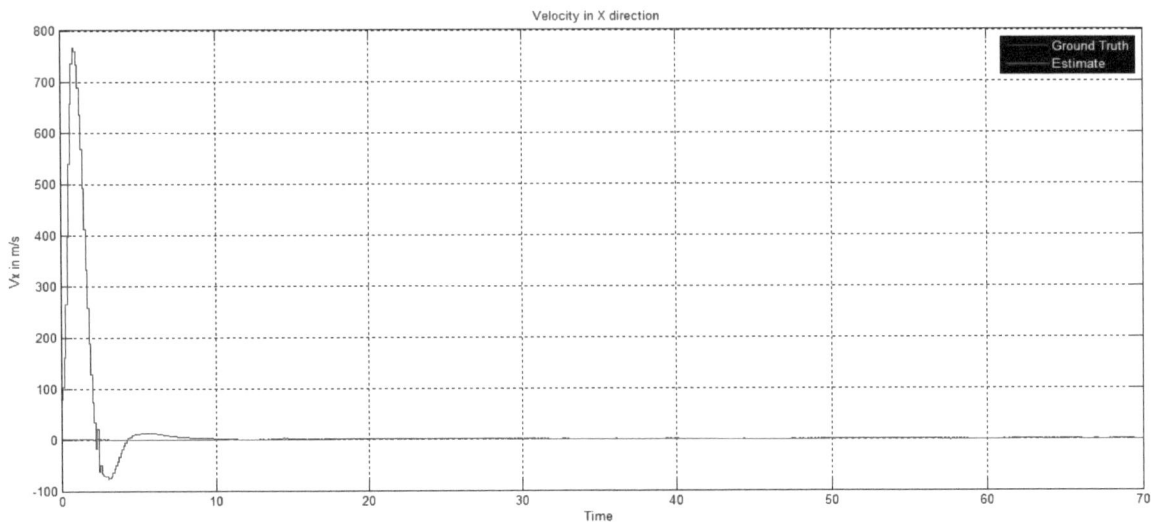

Figure 5.14 Velocity in X direction

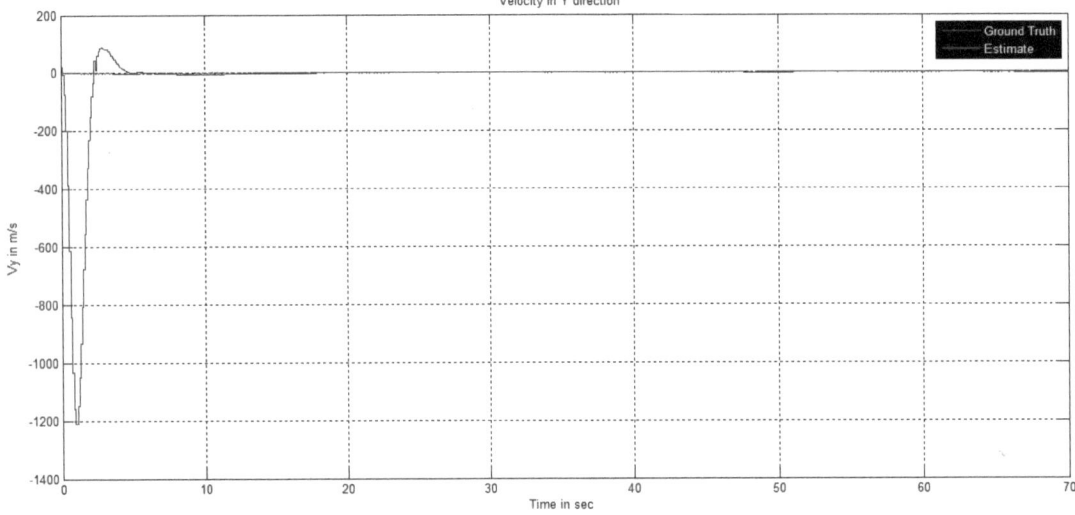

Figure 5.15 Velocity in Y direction

Figure 5.14 and 5.15 provide velocity estimates where differentiated GPS data is used to derive ground truth velocity. The velocity estimates make use of the accelerometer measurements in the earth reference frame. The noisy accelerometer data is smoothened in the filter using suitable Q matrix with higher process noise covariance.

5.2 Position And Velocity Estimates Using 4 State EKF (W/O Accelerometer data)

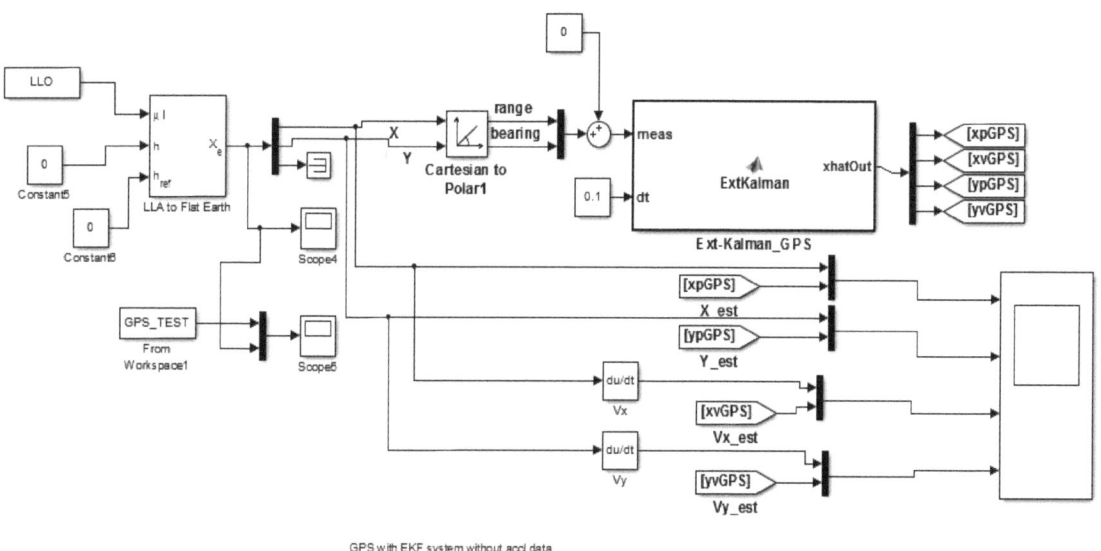

Figure 5.16 EKF without accelerometer data

In the above set up, the equations (5.1) used to estimate the inertial velocity are reduced to a form given below without the accelerometer data. In cases, strictly for simulation purposes, noisy acceleration data are omitted during the estimating process. Note that the velocity model in this case takes the form of a random walk model.

$$(Inertial\ Velocity) \begin{cases} V_{x_{k+1}} = V_{x_k} + a_x^E \partial t \\ V_{y_{k+1}} = V_{y_k} + a_y^E \partial t \end{cases} \quad (5.1)$$
$$\text{Where } a_x^E = a_y^E = 0$$

In this case the linearized equation becomes:

$$\hat{X}_{k+1} = A_k X_k \quad (5.2)$$

A comparison between the tracking ability of the EKF with and without accelerometer data is provided and conclusions are drawn based on the best set up of the EKF.

POSITION ESTIMATES:

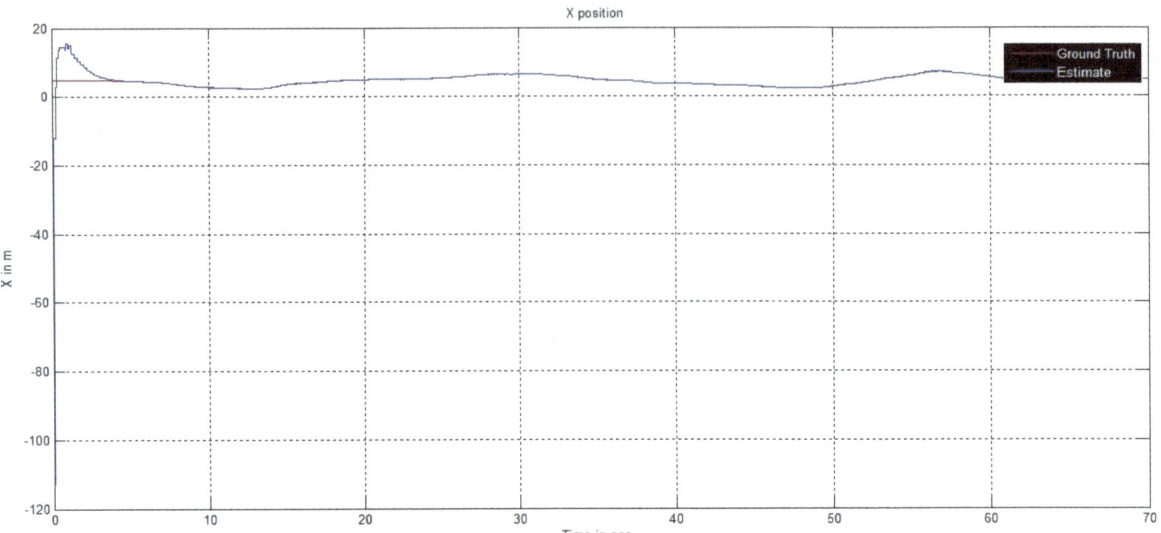

Figure 5.17 X position

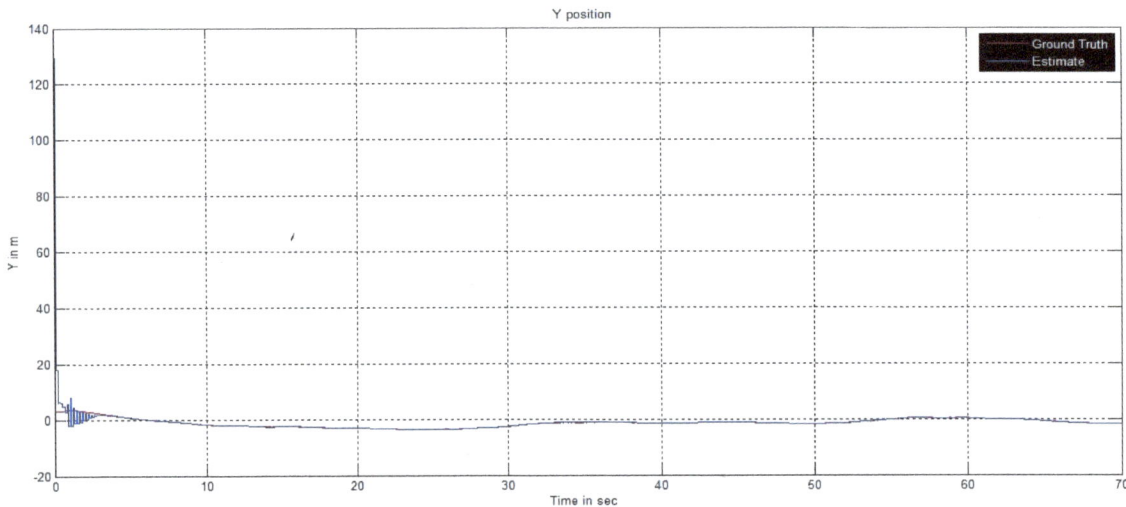

Figure 5.13 Y position

Figures 5.17 and 5.18 compare the X and Y estimates to the ground truth. It can be observed that the EKF performs better along the X axis than the Y axis when compared to respective plots with accelerometer data included.

VELOCITY ESTIMATES:

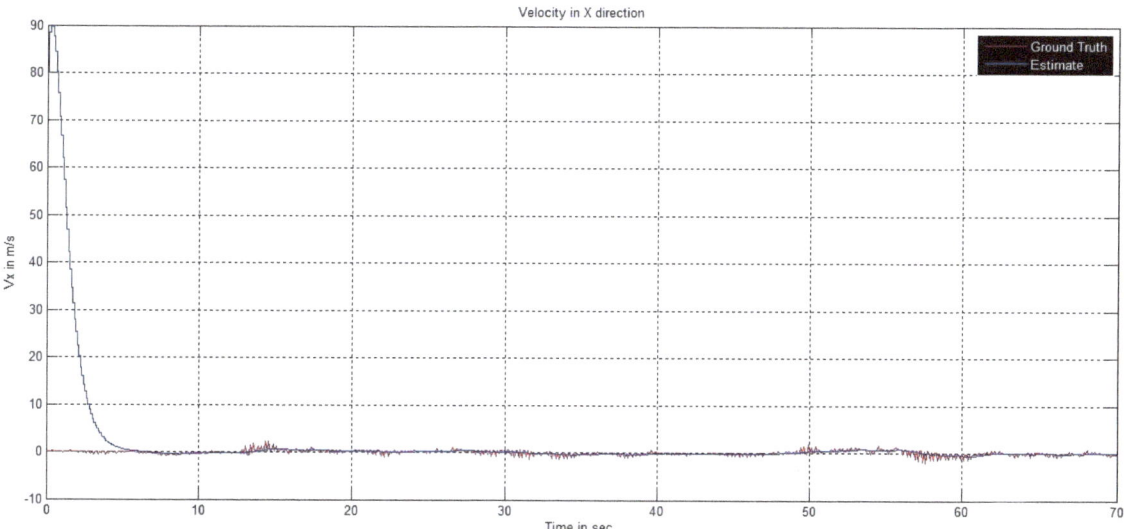

Figure 5.19 Velocity in X direction

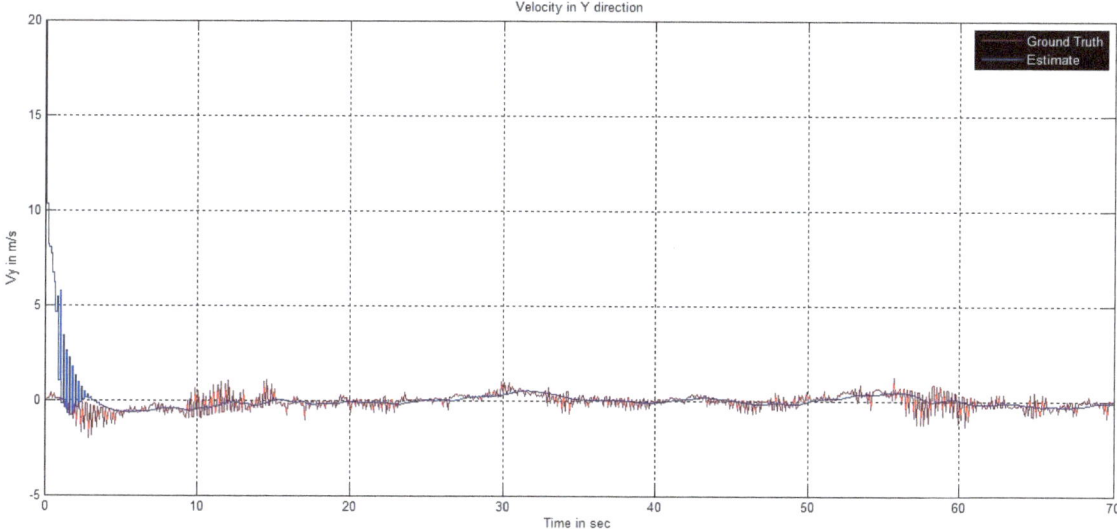

Figure 5.20 Velocity in Y direction

The velocity estimates shown in Figure 5.19 and Figure 5.20 are less robust than the estimates shown earlier with the accelerometer data included.

5.3 4-State Extended Kalman Filter with Additive White Gaussian noise

In figures 5.21-5.24 the position and velocity estimates with three different levels of measurement noise are shown. Each case has a random additive white Gaussian noise augmented to the measurement data separately.

In spite of external noise which can arise due to many factors, the filter performs robustly and tracks the position and velocity states in all the three cases. The three cases where a variance to the range of 5m, 12.5m and 20m are added as AWGN. Similarly the variance to the bearing for the three cases are at 2.5 deg, 5 deg and 8 deg.

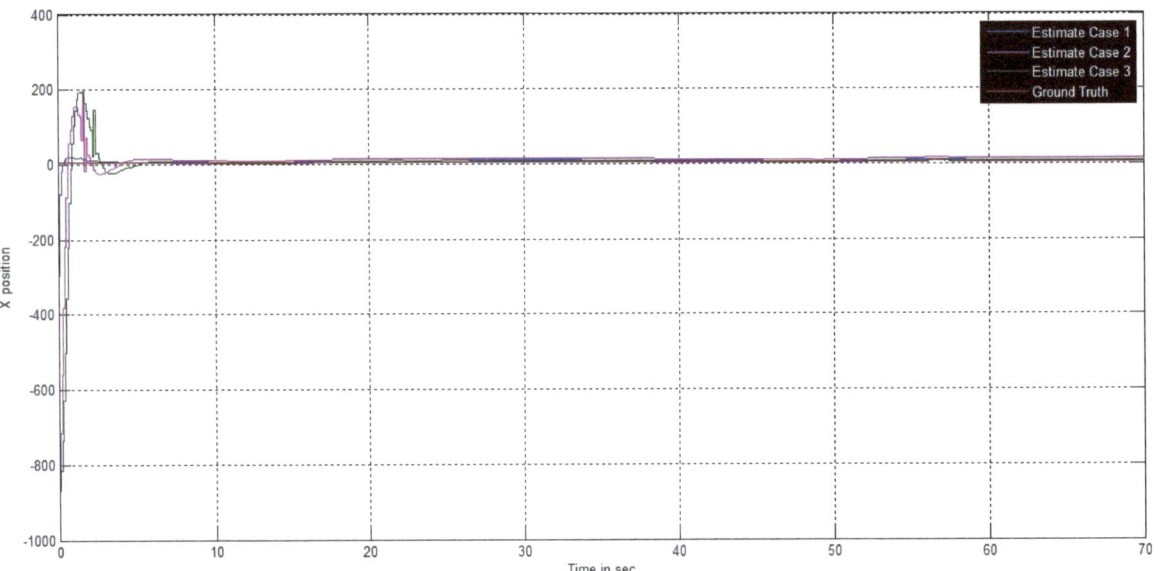

Figure 5.21 X estimate with measurement noise characteristics

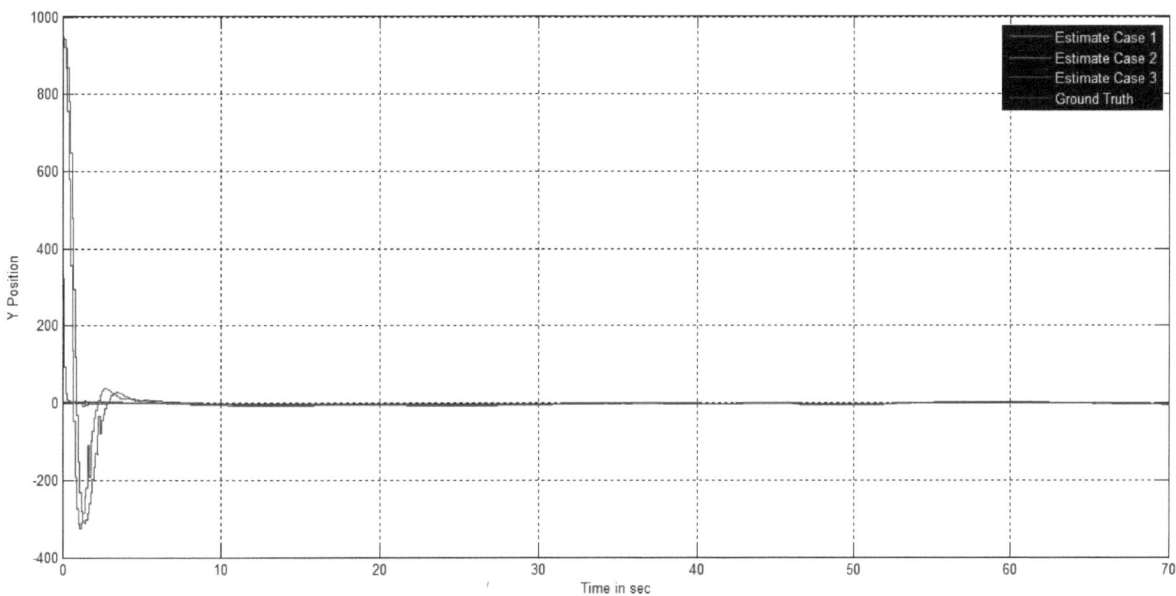

Figure 5.22 Y Position with measurement noise characteristics

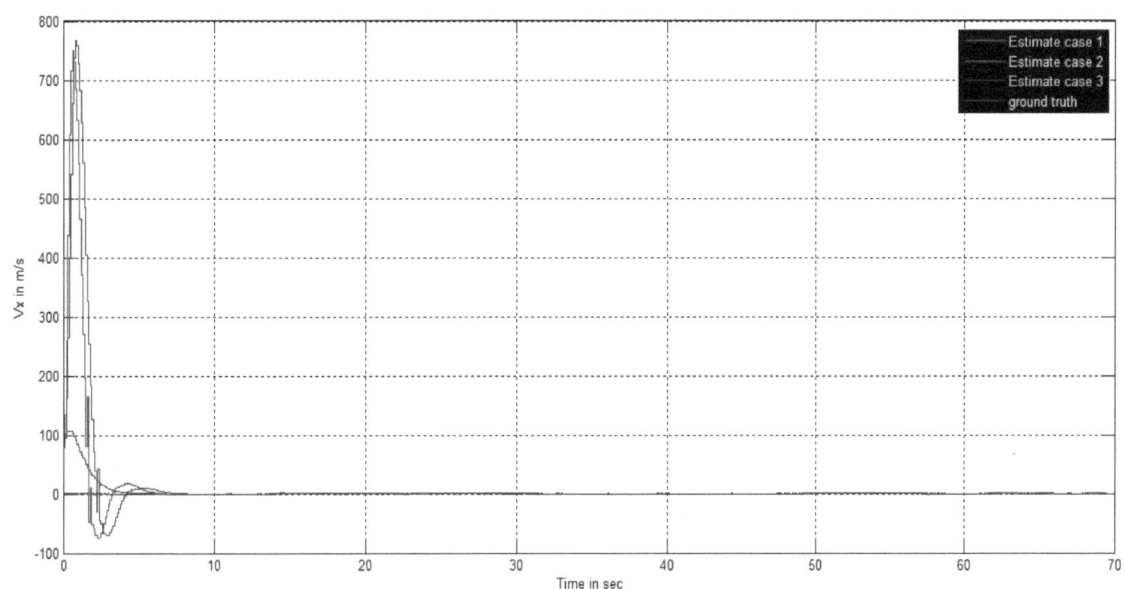

Figure 5.23 Velocity in X direction measurement noise characteristics

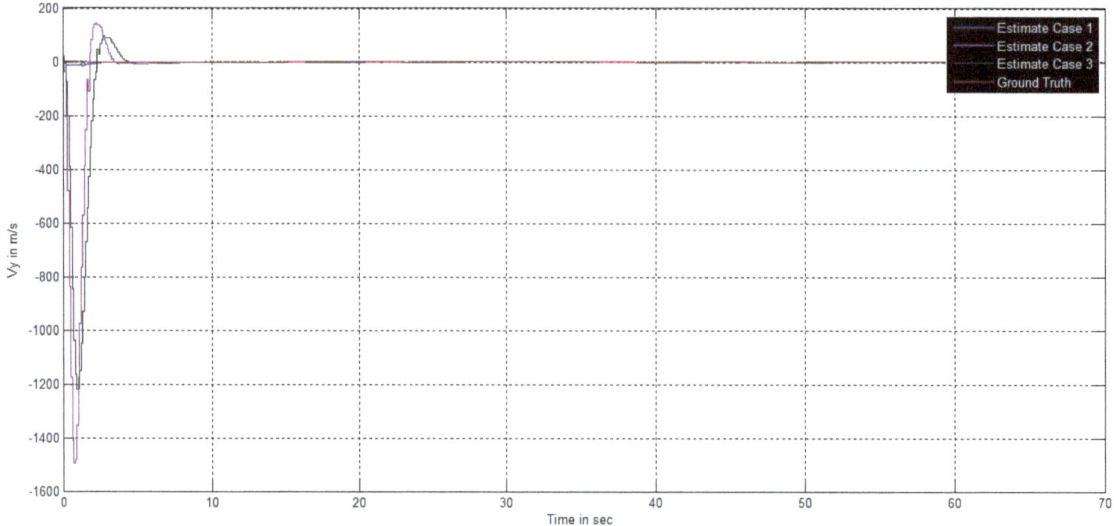

Figure 5.24 Velocity in Y direction measurement noise characteristics

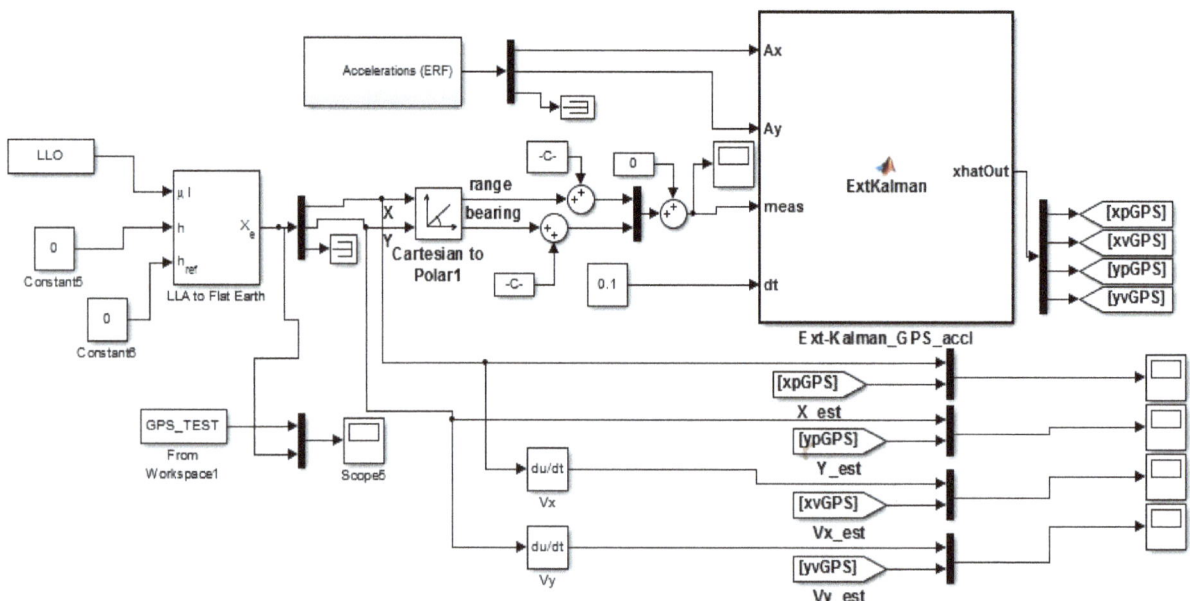

Figure 5.25 Simulink model of EKF with Measurement Noise

5.4 Extended Kalman Filter for tracking Multiple Landmarks/Targets:

The simulation discussed below depicts the navigation or tracking ability of the 4 state EKF when two or more landmarks are introduced in the inertial frame The introduction of multiple landmarks in the environment for vehicle navigation aims at studying the versatility and robustness of the EKF logic which can be experimentally tested in real-time scenarios involving multiple target tracking for reconnaissance or any other defense purposes.

In this simulated example, the first target is at the origin (0, 0) and the second object is at distance of 5m from the first target along the y-axis (0, 5). The logic involves the change in the calculation of the range and the bearing angle. Given that there are two targets at (x_1, y_1) and (x_2, y_2) respectively, the range and bearing to these targets are given by:

$$\begin{aligned} Range_1 &= \sqrt{(x-x_1)^2 + (y-y_1)^2} \\ Bearing_1 &= \tan^{-1}\left(\frac{y-y_1}{x-x_1}\right) \\ Range_2 &= \sqrt{(x-x_2)^2 + (y-y_2)^2} \\ Bearing_2 &= \tan^{-1}\left(\frac{y-y_2}{x-x_2}\right) \end{aligned} \qquad (5.3)$$

Here (x_1, y_1) and (x_2, y_2) represent the co-ordinates of the 2 targets. The EKF tracking with respect to 2 targets, 1 target and the ground truth (raw GPS data) is shown in the following simulations. In later stages, the camera data in the inertial frame would replace the GPS position data. The above equation 5.3 can be implemented as a separate subsystem to perform the Cartesian to Polar transformation based on the X-Y coordinates of the targets and the vehicle in polar plane.

POSITION ESTIMATES:

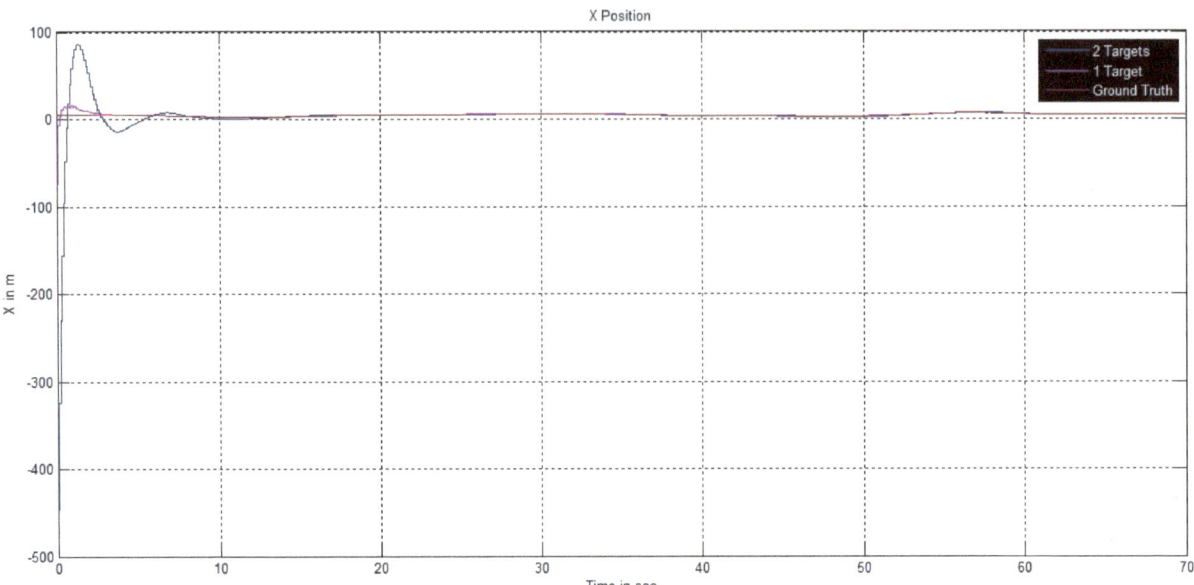

Figure 5.26 X position

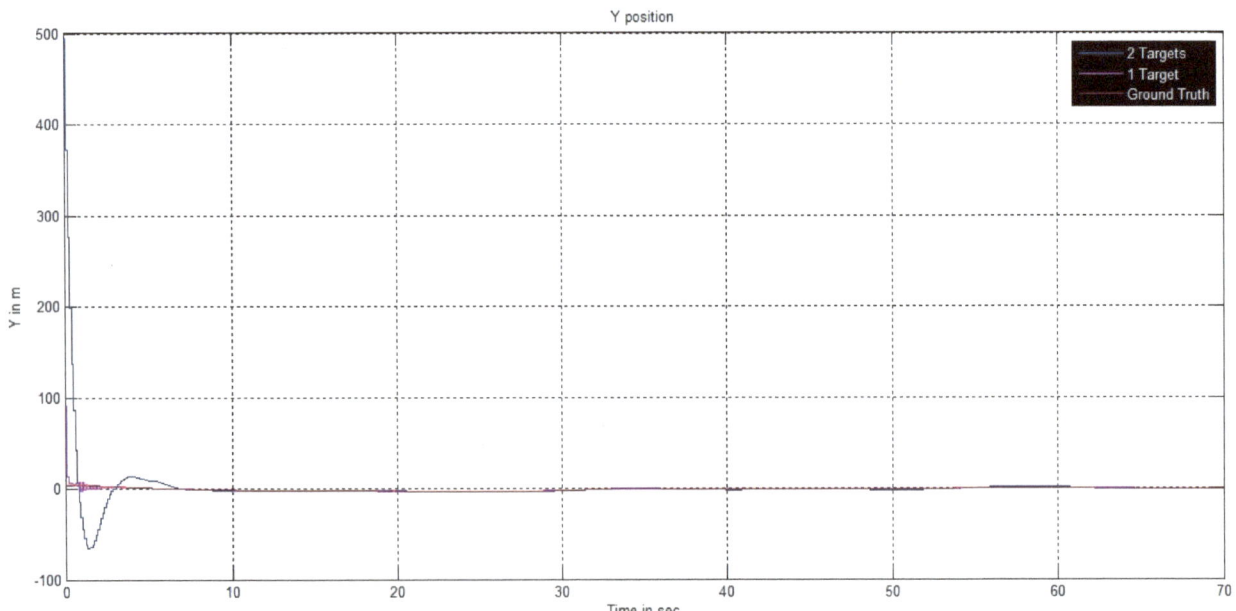

Figure 5.27 Y Position

Figures 5.26 and 5.27 show the performance of the filter for tracking the X-Y position when two targets are present in the environment and when only one target is present.

VELOCITY ESTIMATES:

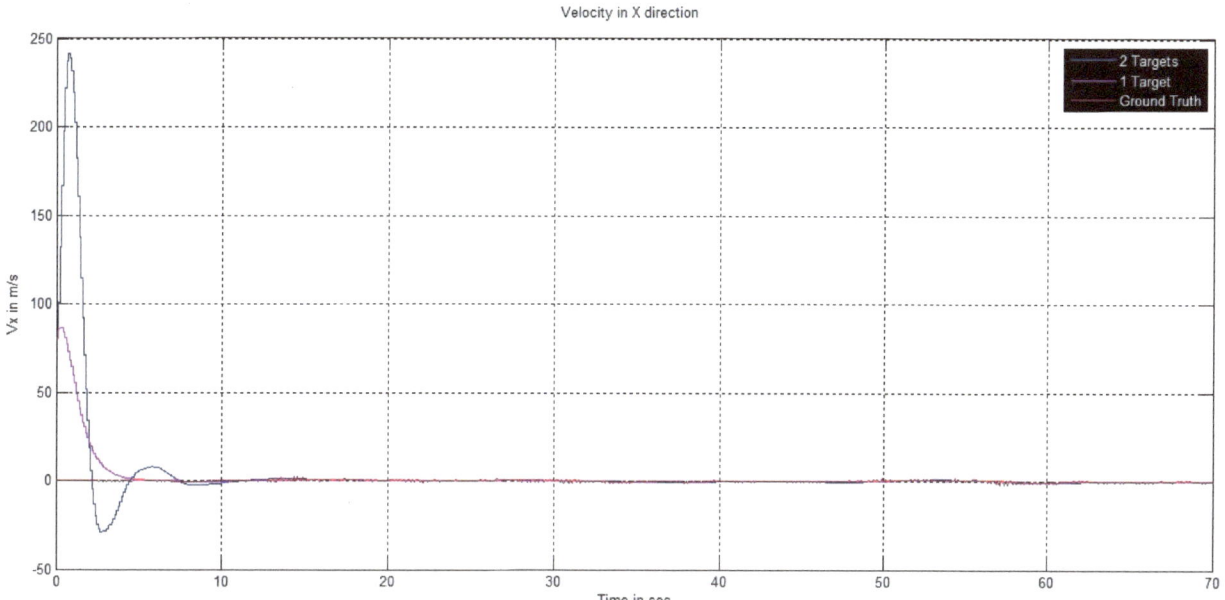

Figure 5.28 Velocity in X direction

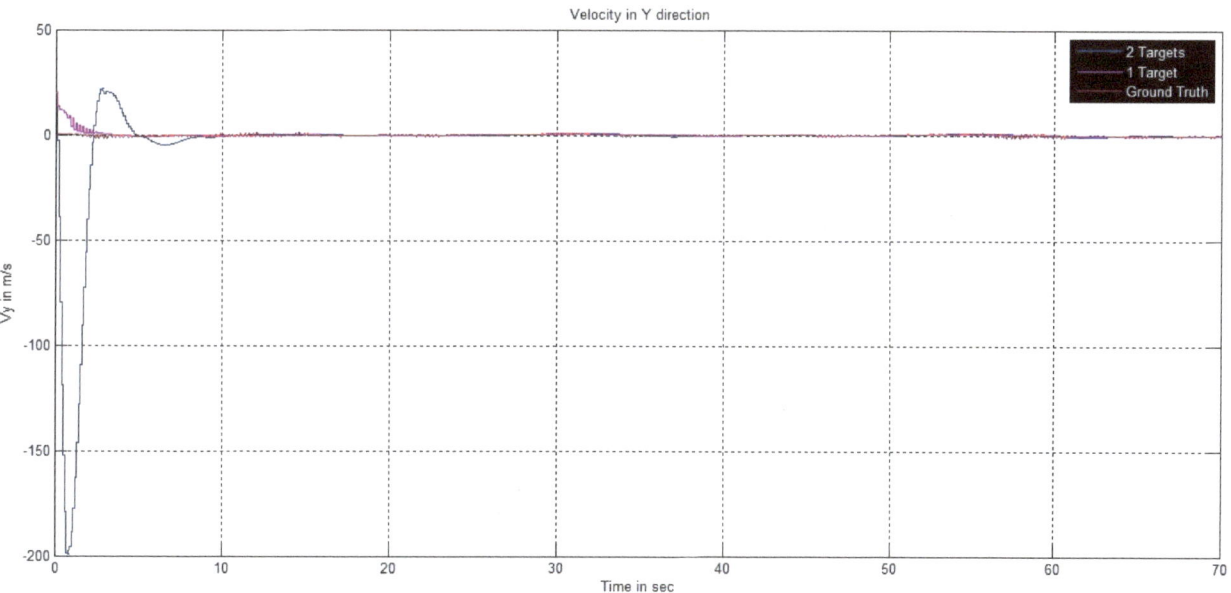

Figure 5.29 Velocity in Y-direction

Figures 5.28 and 5.29 represent the velocity estimation for 2 targets and one target in X-Y direction. The performance of the filter is more robust when tracking the single target. The tracking performance for two or more targets can be improved by placing them at a further distance from each other.

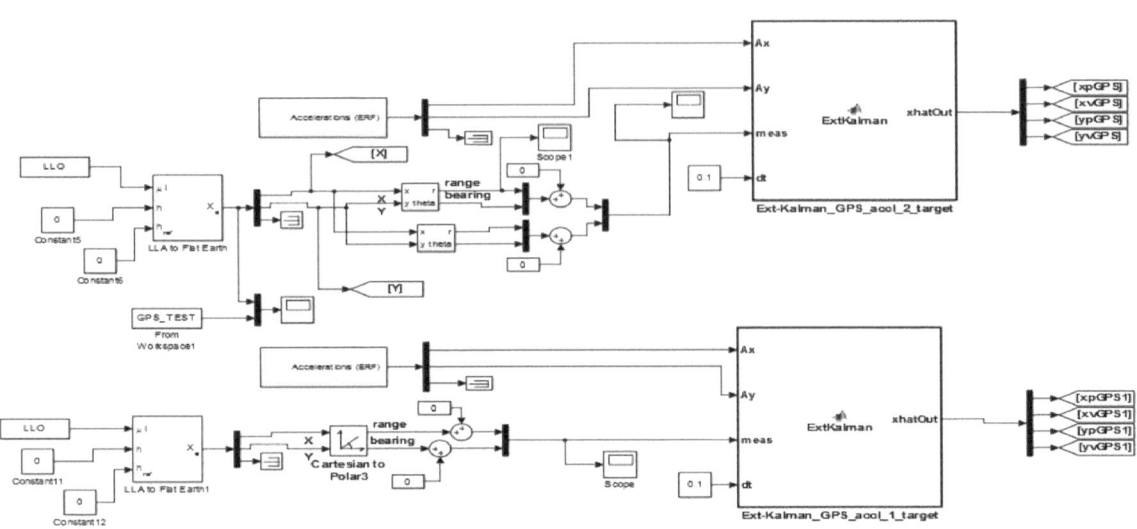

Figure 5.30 Simulink Model for Multiple Target Tracking

Figure 5.30 represents the implementation of the multiple landmark/target tracking example. It is to be noted that the size of the measurement matrix changes in the case of 2 or more targets tracked. The size of the measurement matrix becomes 'n' times the number of targets. Similarly this reflects in the dimensions of measurement Jacobian matrix. The above example can be expanded to track more targets based on the application in which it is used.

5.5 6-State Extended Kalman Filter

A final set of experiments is conducted to estimate the 6 states corresponding to the position and velocity estimates in X-Y-Z directions. This 3D example uses the same data from the MicroStrain attached to the unmanned ground vehicle taking into account the Z direction as well. The equations corresponding to the position and the velocity estimates reflect are as follows:

$$(\text{Inertial Position}) \begin{cases} X_{k+1} = X_k + V_{xk} \partial t \\ Y_{k+1} = Y_k + V_{yk} \partial t \\ Z_{k+1} = Z_k + V_{zk} \partial t \end{cases} \quad (5.4)$$

$$(\text{Inertial Velocity}) \begin{cases} V_{xk+1} = V_{xk} + a_x^E \partial t \\ V_{yk+1} = V_{yk} + a_y^E \partial t \\ V_{zk+1} = V_{zk} + a_z^E \partial t \end{cases} \quad (5.5)$$

The implementation of the above 6-state Extended Kalman Filter with the accelerometer data is shown in Figure 5.31

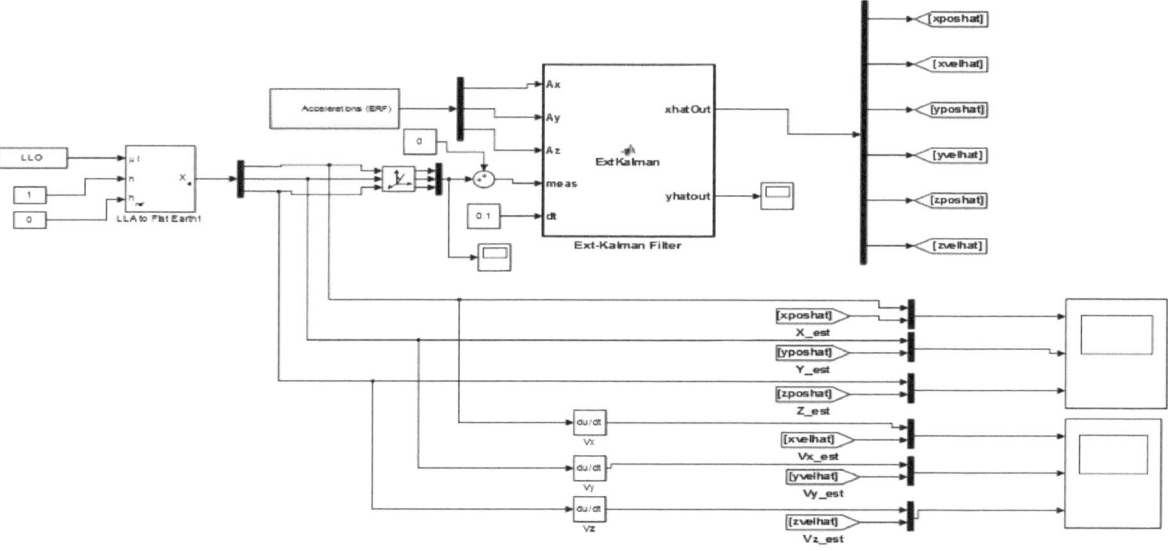

Figure 5.31 Simulink model for 6-state EKF with accelerometer data

The 6-state EKF makes use of a Cartesian to a spherical transformation to estimate the range, bearing angle and zenith angle [75]. The Cartesian to spherical block shown in Figure 5.32 the transformation.

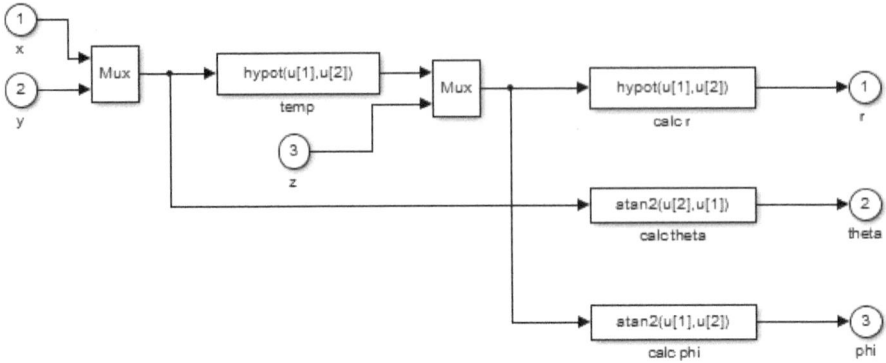

Figure 5.32 Cartesian to Spherical transformation

The state and measurement equations in the filter change accordingly owing to the six state EKF. The equations for the measurements are given in equation (5.6). The Jacobian matrix for the measurement data is obtained by linearizing of these equations about the current state estimate.

$$\begin{aligned} Range(row\,in\,m) &= \sqrt{x^2 + y^2 + z^2} \\ Azimuth(theta\,in\,rad) &= \tan^{-1}\left(\frac{y}{x}\right) \\ Bearing(Phi\,in\,rad) &= \tan^{-1}\left(\frac{\sqrt{x^2+y^2}}{z}\right) \end{aligned} \quad (5.6)$$

In the spherical coordinates, the bearing angle can be represented as the inclination or elevation angle and the azimuth angle is represented as the angle measured from the azimuth reference direction to the orthogonal projection of the line segment connecting the points corresponding to a fixed origin and the vehicle position in n-dimensional space on the reference plane. The spherical co-ordinate transformation can be understood in the Figure 5.33 which depicts the above mentioned equations.

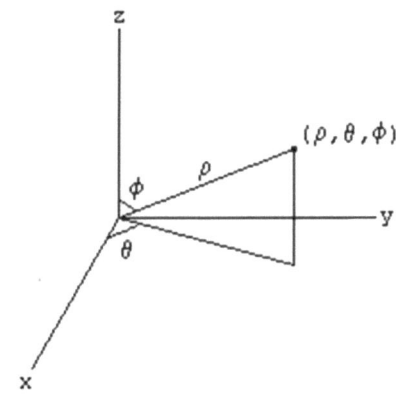

Figure 5.33 Spherical Co-ordinate transformation

The measurement data after the spherical transformation representing the range, bearing and azimuth angle is represented in Figure 5.34. The size of the measurement data matrix varies based on the number of measurements.

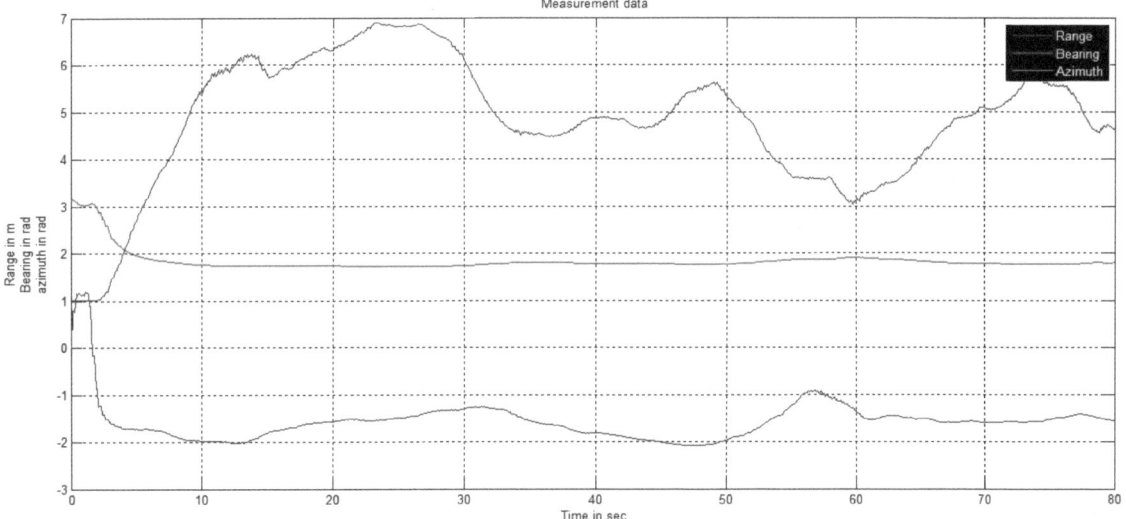

Figure 5.34 Measurement Data

Figure 5.35 and 5.36 represent the 6-state estimates and the ground truth. The position estimates in the Y-Z axis experience a small offset which does not affect the tracking ability in this case. The same application of using multiple objects in the scene can also be incorporated in 3D.

POSITION ESTIMATES:

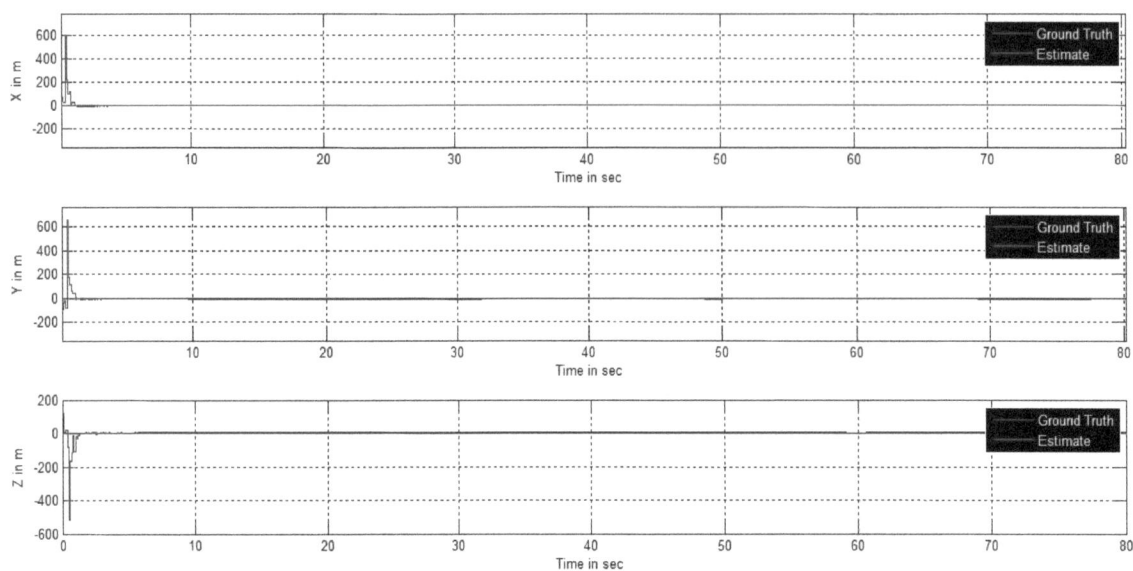

Figure 5.35 Position Estimates for 3D

VELOCITY ESTIMATES:

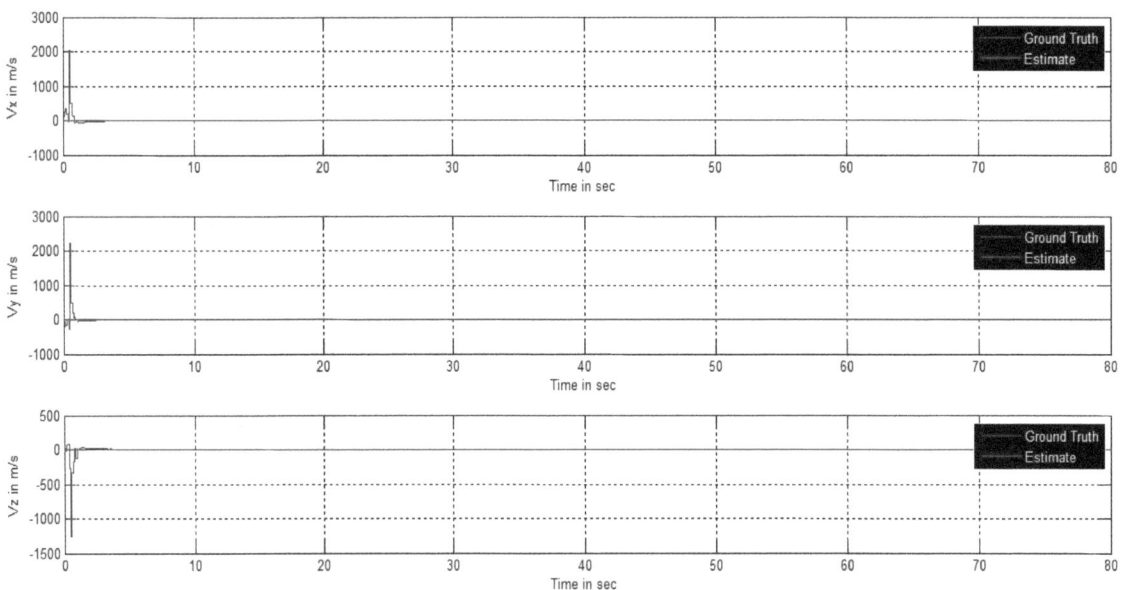

Figure 5.36 Velocity Estimates for 3D

6 THE NEAR FUTURE

The idea of implementing specific image/video processing algorithms for the purpose of vision-aided navigation was discussed in this book. The specific algorithms considered were classified as data independent and data dependent processing algorithms. The CAMshift (continuously adaptive mean shift) algorithm used was a more application-specific algorithm and served the purpose of studying the effect of camera scaling and vehicle orientation on landmark tracking objectives. The ADCOM (advanced compressive tracking) algorithm served the purpose of landmark tracking without focusing on the color probability histogram of the tracked landmark. By design, this algorithm was able to compensate for various types of camera motion, occlusions, background data interference or illumination effects. There is no need to run separate functions to compensate for the above-mentioned effects, especially the effects due to scaling and orientation.

The effectiveness of the two application-specific image processing algorithms was demonstrated with the experimental results, which indicated the tracking of the landmark and estimated the coordinates of the landmark in the camera/image reference frame. The results shown pertain to static landmarks in the environment with the vehicle in motion. The testing of the algorithms using video/image frames collected from high fidelity cameras equipped on an aerial and ground vehicle demonstrate the robustness of the algorithms. It is also to be noted that the CAMshift and ADCOM algorithms are data dependent and data independent respectively, which broadens the scope of using these algorithms for newer reconnaissance tasks.

The comparative study of the two image processing algorithms identified three essential differences applicable to the fundamental operation of these two algorithms. First, the CAMshift algorithm makes use of only the color probability function and tracks the centroid of the landmark while the ADCOM algorithm makes use of features extracted from the landmark being observed. These features are not necessarily restricted to texture or color but are varied over a number of generic parameters.

Second, while using the CAMshift algorithm, owing to the constant adaptability of the bounding box based on the orientation effects, there can be a loss or a delay of the object being tracked as the structure of the tracking object being viewed can become compromised with constant motion of the bounding box. In contrast, the ADCOM algorithm compresses the tracked features into a lower dimensional space without losing any feature data necessary for tracking; hence the structure of the image data is not compromised. Third, the CAMshift algorithm uses a kernel estimation as a classifier but the ADCOM algorithm makes use of a naïve Bayes classifier.

A robust extended Kalman filter (EKF) is proposed for the process of vision-aided navigation in GPS-denied environments. The navigation filter provides the essential solution to the problem of navigation using only a few states and noisy target measurements. This filter was tested using simulated target measurements where the coordinate transformation inside the simulation enables the calculation of the range between the vehicle and the landmark along with the angle of inclination to the landmark, represented in terms of bearing angle. The extended Kalman filter was tested for varying levels of measurement noise, specified in terms of a Gaussian white noise model. This modeled the effect of external noise during real-time implementation on a vehicle and how the filter robustly compensates for the noise disturbances. Data collected from an unmanned ground vehicle provided test data for the Kalman filter as the vehicle's path and Euler angles were calculated for filter processing. A Cartesian to spherical coordinate transformation was used to simulate the range and angles to the target.

The essential feature of the 4-state EKF was the ability to estimate the planar position and velocity of the vehicle with the accelerometer data as input measurements and measurements of multiple landmarks in the environment. The use of such filter logic is well suited for applications where multiple static objects are tracked by the same vehicle for reconnaissance purposes in the defense sector.

A final set of simulations involved the use of a 6-state EKF to estimate the 3D motion of the vehicle while tracking an object at a specific location in the environment. This proved slightly less robust than the 4-state filter as the addition of data in the Z-direction invoked high noise levels, resulting in drift in the state estimates.

6.1 Future Recommendations

The two core processes investigated in this thesis, image processing and navigation filtering, need to be integrated into one vision-aided navigation system. Several components are required to form such an integrated system. One required component is a camera calibration process to derive the transformation of the pixel points in the image reference frame to the inertial (earth) reference frame. Some earlier works have conducted extensive research on this transformation in simulated environments but have failed to test it with high end application specific image processing algorithms with navigation filters [76][77][78]. The pixel points in the inertial frame can be used as a surrogate to the GPS measurements when encountering a GPS-denied terrain or in the case where GPS is accessible. The same data can then be used to navigate to the given target locations calculated using the above camera frame transformation.

Figure 6.1 Vision-aided navigation applications

The application specific algorithms coupled with a 6-state or a simple 4-state EKF can be put to use in real time vision-aided navigation systems such as in the ambulance drone (Flying Defibrillator), door-to-door product delivery, remote sensing, IPMC (Ionic Polymer Metal Composite) powered robotic fish and in agricultural surveillance coupled with a bio-sensor mechanism in the commercial sector. The use of the above experimentally verified techniques in the defense sector expands from reconnaissance applications, tracking multiple moving targets, and augmenting LIDAR imaging to develop a 3D map of an inaccessible location.

The need for newer application specific algorithms has raised the awareness amongst researchers to investigate the methods that are suitable, compatible and robust for usage in real time applications. The intriguing task of providing robust navigation solutions for an unmanned vehicle in environments where GPS is inaccessible, using landmark feature identification techniques lays the foundation for exploring the multifarious domain of UAV navigation in remote, inaccessible and undiscovered terrains. Henceforth, this continued effort pushes the boundary of such advanced and in-depth application-driven research to outer space objectives.

7 BIBLIOGRPAHY

[1] Valavanis, K., 'Advances in unmanned aerial vehicles: state of the art and the road to autonomy', Intelligent Systems, Control and Automation: Science and Engineering 33, 2007.

[2] Bejar, M., Ollero, A., Cuesta, F., 'Modeling and control of autonomous helicopters, advances in control theory and applications',Lect. Notes Control Inf. Sci. 353,2007.

[3] Lee, D., Jin Kim, H., Sastry, S., 'Feedback linearization vs. adaptive sliding mode control for a quadrotor helicopter',Int. J. Control Autom. Syst. 7(3), 419–428,2009.

[4] Bernard, M., Kondak, K., Hommel, G., 'Framework for development and test of embedded flight control software for autonomous small size helicopters',Embedded Systems – Modeling, Technology, and Applications, pp. 159–168,2006.

[5] Monteriù, A., Asthana, P., Valavanis, K., Longhi, S., 'Model-based sensor fault detection and isolation system for unmanned ground vehicles: theoretical aspects (partiandii)',Proceedings of the IEEE International Conference on Robotics and Automation (ICRA),2007.

[6] Conte, G., Doherty, P., 'An integrated UAV navigation system based on aerial image matching', IEEE Aerospace Conference, pp. 1–10, 2008.

[7] Luo P., Pei, H., 'An autonomous helicopter with vision based navigation', IEEE International Conference on Control and Automation, 2007.

[8] R. He, S. Prentice, and N. Roy, 'Planning in Information Space for a Quadrotor Helicopter in a GPS-denied Environment', IEEE Intl. Conf. Robotics and Automation, Pasadena, California, May 2008

[9] He, Z., Iyer, R.V., Chandler, P.R., 'Vision-based UAV flight control and obstacle avoidance', American Control Conference, 2006.

[10] Mondragon, I.F., Campoy, P., Correa, J.F., Mejias, L.: 'Visual model feature tracking for UAV control', IEEE International Symposium on Intelligent Signal Processing, WISP, 2007.

[11] Campoy, P., Correa, J.F., Mondragón, I., Martínez, C., Olivares, M., Mejías, L., Artieda, J.: 'Computer vision onboard UAVs for civilian tasks'. J. Intell. Robot. Syst. 54(1–3), 105–135 ,2009.

[12] Chowdhary,Girish., Johnson Eric N.,Magree Daniel., Wu Allen., Shein Andy.,: 'GPS-Denied Indoor and Outdoor Monocular Vision Aided Navigation and Control of Unmanned Aircraft',January,2013.

[13] Caballero, F., Merino, L., Ferruz, J., Ollero, A., 'Vision-based odometry and SLAM for medium and high altitude flying UAVs'. J. Intell. Robot. Syst. 54(1–3), 137–161,2009.

[14] Williams, Paul., Crump, Michael., 'All-Source Navigation For Enhancing UAV Operations in GPS-Denied Environments',28th International Congress of Aeronautical Sciences,2012.

[15] Smith, R.C., On the representation of spatial uncertainty. Int. J. Robotics Research, 5(4), pp.56-68,1987.

[16] Merz,T.,Duranti,S.,Conte,G. 'Autonomous landing of an unmanned helicopter based on vision and inertial sensing',Experimental Robotics IX, Springer Tracts in Advanced Robotics, vol. 21, pp. 343–352,2006.

[17] Meingast, M., Geyer, C., Sastry, S., 'Vision based terrain recovery for landing unmanned aerial vehicles',43rd IEEE Conference on Decision and Control (CDC),vol.2,pp.1670–1675,2004.

[18] Shakernia, O., Vidal, R., Sharp, C.S., Ma, Y., Sastry, S.S.: Multiple view motion estimation and control for landing an unmanned aerial vehicle, Proceedings of the IEEE International Conference on Robotics and Automation (ICRA), pp. 2793–2798 (2002)

[19] Saripalli, S., Montgomery, J., Sukhatme, G., 'Visually-guided landing of an unmanned aerial vehicle',IEEE Trans. Robot. Autom. 19(3), 371–381,2003.

[20] Saripalli, S., Sukhatme, G.S., 'Landing a helicopter on a moving target', IEEE International Conference on Robotics and Automation (ICRA), pp. 2030–2035,2007.

[21] Garcia-Padro, P.J., Sukhatme, G.S., Montgomery, J.F., 'Towards vision-based safe landing for an autonomous helicopter', Robotics and Autonomous Systems, vol. 38, no. 1, pp. 19–29(11). Elsevier, 2002.

[22] Johnson, A., Montgomery, J., Matthies, L., 'Vision guided landing of an autonomous helicopter in hazardous terrain',Proceedings of the IEEE International Conference on Robotics and Automation,2005.

[23] Templeton, T., Shim, D.H., Geyer, C., Sastry, S., 'Autonomous vision-based landing and terrain mapping using am MPC-controlled unmanned rotorcraft', Proceedings of the IEEE International Conference on Robotics and Automation, pp. 1349–1356, 2007.

[24] W. Kropatsch, 'History of Computer Vision A Personal Perspective', Institute of Computer Aided Automation 183/2 Vienna University of Technology Pattern Recognition and Image Processing Group, 2008.

[25] Z. Wang and F. Yang, 'Object Tracking Algorithm Based on Camshift and Grey Prediction Model in Occlusions', in the 2nd International Conference on Computer Application and System Modeling (2012), France, 2012.

[26] Fukunaga, K. (1990), 'Introduction to Statistical Pattern Recognition', 2nd Edition. Academic Press, New York, 1990.

[27] D. Comaniciu, V. Ramesh, and P. Meer, 'Real-time tracking of non-rigid objects using mean shift', IEEE Proc. on Computer Vision and Pattern Recognition on, pages673–678, 2000.

[28] R. Collins, 'Mean-shift Tracking', Computer Science Engineering,CSE598G, Penn State University, 2006.

[29] L. Jae-Yeong and W. Yu, 'Visual tracking by partition-based histogram backprojection and maximum support criteria', in Robotics and Biomimetics (ROBIO), 2011 IEEE International Conference, Karon Beach, Phuket, 2011, pp. 2860 - 2865.

[30] Schugk, D., Kummert, A., and Nunn, C., 'Adaptation of the Mean Shift Tracking Algorithm to Monochrome Vision Systems for Pedestrian Tracking Based on HoG-Features', SAE Technical Paper 2014-01-0170, 2014.

[31] Bradski, G. R. (1998), 'Computer vision face tracking for use in a perceptual user interface', Intel Technology Journal, 2nd Quarter, 1998.

[32] Freeman, W. T., Tanaka, K., Ohta, J. and Kyuma, 'Computer Vision for Computer Games', Int. Conf. On Automatic Face and Gesture Recognition, pp.100-105, 1996.

[33] P. Fieguth and D. Terzopoulos, 'Color-based tracking of heads and other mobile objects at video frame rates,' In Proc. Of IEEE CVPR, pp. 21-27, 1997.

[34] Hu J., Juan C., Wang J.: 'A spatial-color mean-shift object tracking algorithm with scale and orientation estimation', Pattern Recognition Letters, 2008, 29, (16), pp. 2165-2173.

[35] M. Hunke and A. Waibel, 'Face locating and tracking for human-computer interaction,' Proc. Of the 2gth Asilomar Conf. On Signals, Sys. and Comp., pp. 1277- 128 I, 1994

[36] K. Sobottka and I. Pitas, 'Segmentation and tracking of faces in color images,' Proc. Of the Second Intl. Conf. On Auto. Face and Gesture Recognition, pp. 236-241, 1996

[37] M. Swain and D. Ballard, 'Color indexing,' Intl. J. of Computer Vision, 7(I) pp. 1 1-32, 1991.

[38] Horn R. A., Johnson C. R., "Topics in Matrix Analysis", Cambridge University Press, U.K., 1991.

[39] Volkan Cevher, Aswin Sankaranarayanan, Marco F. Duarte1, Dikpal Reddy, Richard G. Baraniuk, and Rama Chellappa, 'Compressive sensing for Background subtraction', Rice University,ECE,Houston TX 77005,2008.

[40] Aggarwal, A., Biswas, S., Singh, S., Sural, S., Majumdar, A.K, 'Object Tracking Using Background Subtraction and Motion Estimation in MPEG Videos'. In: ACCV, Springer (2006) 121–130

[41]Cao, Y., Lei, Z., Huang, X., Zhang, Z. and Zhong, T, 'A Vehicle Detection Algorithm Based on Compressive Sensing and Background Subtraction", AASRI Procedia, 1, pp.480-485,2012.

[42] Fan, B., Du, Y. and Cong, Y. (2014). 'Online Learning Discriminative Dictionary with Label Information for Robust Object Tracking', Abstract and Applied Analysis, 2014, pp.1-12.

[43]Wu, Y., Jia, N. and Sun, J., 'Real-time multi-scale tracking based on compressive sensing',2014.

[44] Garrett Warnell and Rama Chellappa, 'Compressive Sensing in Visual Tracking, Recent Developments in Video Surveillance', Dr. Hazem El-Alfy (Ed.), ISBN: 978-953-51-0468-1, 2012.

[45] K. Zhang, L. Zhang and M. Yang, 'Real-Time Compressive Tracking', ECCV 2012, Part III, LNCS 7574, pp. pp. 866–879, 2012.

[46] P. Frankl and H. Maehara, 'The Johnson-Lindenstrauss lemma and the sphericity of some graphs', Journal of Combinatorial Theory A, 44(3):355–362, 1987.

[47] P. Li, T. Hastie and k. Church, 'Very Sparse random projections', in Proceedings of the 12th ACM SIGKDD international conference on Knowledge discovery and data mining, New York, 2006.

[48] E. Candès, 'The restricted isometry property and its implications for compressed sensing',Comptes Rendus Mathematique, vol. 346, no. 9-10, pp. 589-592, 2008.

[49] Li, H., Shen, C., Shi, Q., 'Real-time visual tracking using compressive sensing', In: CVPR, pp. 1305–1312, 2011.

[50] Wright, J., Yang, A., Ganesh, A., Sastry, S., Ma, Y, 'Robust face recognition via sparse representation', PAMI 31, 210–227, 2009.

[51] Liu, L., Fieguth, P., 'Texture classification from random features', PAMI 34, 574–586, 2012.

[52] Dimitris Achlioptas, 'Database-friendly random projections: Johnson-Lindenstrauss with binary coins', Journal of Computer and System Sciences, 66(4):671–687, 2003.

[53] Ella Bingham and Heikki Mannila. Random projection in dimensionality reduction: Applications to image and text data. In Proc. of KDD, pages 245–250, San Francisco, CA, 2001.

[54] K. Wu and X. Guo, 'Compressive Sensing with Sparse Measurement Matrices', in Vehicular Technology Conference, China, pp. pp 1-5, 2011.

[55] A. Panning, A. Al-Hamadi, R. Niese and B. Michaelis, 'Facial expression recognition based on Haar-like feature detection', Pattern Recognition and Image Analysis, vol. 18, no. 3, pp. 447-452, 2008.

[56] R. Lienhart and J. Maydt, 'An Extended Set of Haar-like Features for Rapid Object Detection', in image Processing Proceedings International Conference, USA, pp. I-900 - I-903 vol.1, 2002.

[57] Yang, J., Bouzerdoum, A., Tivive, F. & Phung, S., 'Dimensionality reduction using compressed sensing and its application to a large-scale visual recognition task', WCCI 2010 IEEE World Congress on Computational Intelligence, pp. 1607-1614, 2010.

[58] H. Zhang, 'EXPLORING CONDITIONS FOR THE OPTIMALITY OF NAÏVE BAYES', International Journal of Pattern Recognition and Artificial Intelligence, vol. 19, no. 02, pp. 183-198, 2005.

[59] J. Xue and D. Titterington, 'Comment on "On Discriminative vs. Generative Classifiers: A Comparison of Logistic Regression and Naive Bayes"', Neural Processing Letters, vol. 28, no. 3, pp. 169-187, 2008.

[60] P. Diaconis and D. Freedman, 'Asymptotics of Graphical Projection Pursuit', The Annals of Statistics, vol. 12, no. 3, pp. 793-815, 1984.

[61] V. Indelman, P. Gurfil, E. Rivlin and H. Rotstein, 'Real-Time Vision-Aided Localization and Navigation Based on Three-View Geometry', IEEE Trans. Aerosp. Electron. Syst., vol. 48, no. 3, pp. 2239-2259, 2012.

[62] Greg Welch, Gary Bishop, 'An Introduction to the Kalman Filter', University of North Carolina at Chapel Hill Department of Computer Science, 2001.

[63] M.S.Grewal, A.P. Andrews, 'Kalman Filtering - Theory and Practice Using MATLAB', Wiley, 2001.

[64] Maybeck, P. S. 'The Kalman filter: An introduction to concepts', Autonomous Robot Vehicles. I. J. Cox and G. T. Wilfong. New York, Springer-Verlag: 194-204, 1990.

[65] Gary Bishop and Greg Welch, 'An Introduction to the Kalman Filter', University of North Carolina SIGGRAPH 2001 course notes. ACM Inc., North Carolina, 2001.

[66] V. Sazdovski, T. Kolemishevska-Gugulovska and M. Stankovski, 'Kalman Filter Implementation For Unmanned Aerial Vehicles Navigation Developed Within A Graduate Course', Institute of ASE at Faculty of EE, St. Cyril and Methodius University, MK-1000, Skopje, Republic of Macedonia, 2005.

[67] John Spletzer, 'The Discrete Kalman Filter'. Lecture notes CSC398/498. Lehigh University. Bethlehem, PA, USA. March 2005.

[68] Rudolph van der Merwe, Alex T. Nelson, Eric Wan, 'An Introduction to Kalman Filtering.' OGI School of Science & Engineering lecture. Oregon Health & Sciences University. November 2004.

[69] Simon Julier and Jeffrey Uhlmann, 'A new extension of the kalman filter to nonlinear systems', Int. Symp. Aerospace/Defense Sensing, Simul. And Controls, Orlando, FL, 1997.

[70] N.J. Gordon, D.J. Salmond, and A.F.M. Smith., 'A novel approach to nonlinear/non-Gaussian Bayesian state estimation', IEEE Proceedings on Radar and Signal Processing, volume 140, pages 107-113, 1993.

[71] G. Consulting, 'Using an Extended Kalman Filter for Object Tracking in Simulink', Goddardconsulting.ca, 2014. [Online]. Available: http://www.goddardconsulting.ca/simulink-extended-kalman-filter-tracking.html#Figure2.

[72] Robotics.coroware.com, 'CoroBot Classic four wheel drive (CL4)', 2014. [Online]. Available: http://robotics.coroware.com/Template2.aspx?WebPage=CL4.

[73] Microstrain.com, '3DM-GX3®-45 -- Product no longer stocked – limited availability', 2011. [Online]. Available: http://www.microstrain.com/inertial/3dm-gx3-45.

[74] Shop.gopro.com, 'GoPro Official Website: The World's Most Versatile Camera', 2014. [Online]. Available: http://shop.gopro.com/.

[75] Zwillinger, D. (Ed.). 'Spherical Coordinates in Space.' §4.9.3 in CRC Standard Mathematical Tables and Formulae. Boca Raton, FL: CRC Press, pp. 297-298, 1995.

[76] A. E. Johnson and L. H. Matthies, "Precise image-based motion Estimation for autonomous small body exploration," in Proc. 5th Int'l Symp. On Artificial Intelligence, Robotics and Automation in Space, Noordwijk, The Netherlands, June 1-3 1999, pp. 627–634

[77] J. Lobo and J. Dias, 'Relative Pose Calibration Between Visual and Inertial Sensors', The International Journal of Robotics Research, vol. 26, no. 6, pp. 561-575, 2007. [78] Kelly and G. Sukhatme, 'Fast relative pose calibration for visual and inertial sensors', Springer Berlin Heidelberg, vol. 54, pp. 515-524, 2009.

ABOUT THE AUTHOR

Tennyson Samuel John is a highly self-motivated professional actively involved in designing, testing, modelling multifarious systems in the aerospace industry and contributing continually to the industry's growth. He is a budding entrepreneur who constantly dreams of making a difference in the world through fascinating technology. He has a heart of an engineer and a brain of a businessman.

www.ingramcontent.com/pod-product-compliance
Lightning Source LLC
Chambersburg PA
CBHW051020180526
45172CB00002B/420